TRUE
NORTH
SERIES

Your Guide to **Zion**
AND
Bryce
Canyon
NATIONAL
PARKS

A DIFFERENT PERSPECTIVE

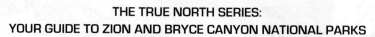

THE TRUE NORTH SERIES:
YOUR GUIDE TO ZION AND BRYCE CANYON NATIONAL PARKS

Second printing: October 2019

ISBN-13: 978-0-89051-580-8
ISBN-10: 0-89051-580-8
Library of Congress Number: 2010922021

Unless otherwise noted, all Scripture is from the New American Standard Bible.

Cover and interior design by Jennifer Bauer and Tom Vail.

Printed in China

For other great titles, please visit the Master Books website at www.masterbooks.com or the Creation Research Society website at www.creationresearch.org.

For information regarding author interviews, please contact the publicity department at (870) 438-5288.

The Narrows, Zion

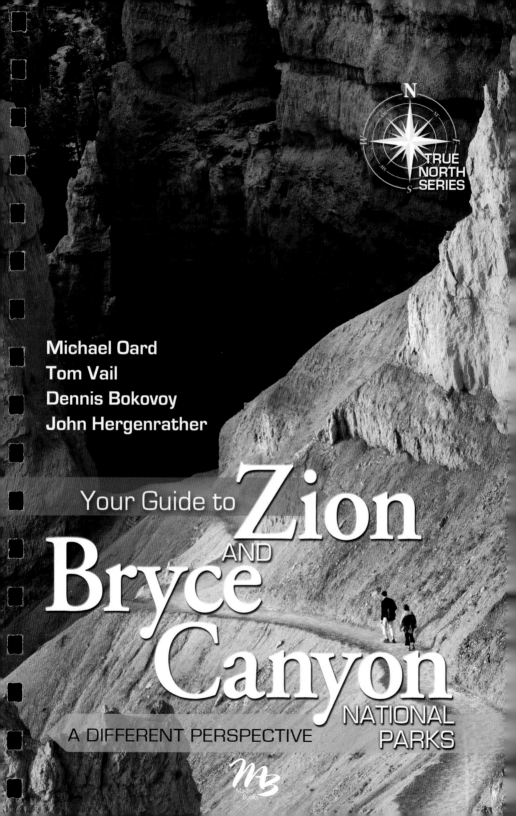

TRUE
NORTH
SERIES

Michael Oard
Tom Vail
Dennis Bokovoy
John Hergenrather

Your Guide to *Zion*
AND
Bryce
Canyon
NATIONAL
PARKS

A DIFFERENT PERSPECTIVE

Master Books

Bryce Canyon National Park

Some geologic sites are absolutely stunning. They thrill the viewer from the inside out! From the very first breath-taking view, you're captivated. Your camera just can't take pictures fast enough. They never grow old, and keep calling you back.

So it is with Zion Canyon and Bryce Canyon. From the multi-colored formations to the abundant trees and wildlife, your senses can hardly take it all in. The artistic masterpiece on display looks like the purposeful work of a Great Master. Yet we are told that it's all due to long ages of deposition and erosion. How can something so beautiful be the result of destructive natural forces? Is there a better explanation?

We read in Scripture that the God of Creation formed all things to be "very good" (Genesis 1:31) in the beginning.

Surely Adam and Eve likewise reveled in the beauty surrounding them. Did they see what we see? Or was it even better? Scripture also tells us of a time when the original perfection was distorted due to man's actions, and then fully restructured by global tectonic and hydraulic cataclysmic forces. The results of such overwhelming forces include extensive erosion and re-deposition, massive uplifting, and major folding and faulting of rock. Everything is now different. If the great Flood of Noah's day was really global in extent, where could you go on planet Earth and not see a flood-caused terrain? If the earth has been flooded, and all we see is the result of such a flood, then what happens if we don't include that flood in our thinking? It means we don't have a clue. How can we arrive at truth if we deny truth at the start?

This handy guide to popular landmarks does us all a great service, for it not only helps us get to the proper sites, but explains what we see, and helps us interpret the rocks and other features in a proper context. The authors are well-experienced in geologic observation as well as biblical application. They can guide us to the places where we can make the best observations and help usher us to satisfying interpretations. They'll not offend the reader with illogical reasoning, nor will they dogmatically insist on a narrow conclusion. Prepare for an uplifting and exhilarating experience, as they help you see the sites and derive your own conclusions, based on the real data.

One more point to ponder. If this world is the destroyed remnant of the once "very good" world, think how beautiful that world must have been before its destruction to still bear such beauty. And then recognize that Scripture tells of a coming "new heavens and a new Earth, in which righteousness dwells" (2 Peter 3:13) once again, with both creation's purpose fulfilled and man's fellowship with the Creator restored.

John D. Morris, PhD
President of the
Institute for
Creation Research

Zion National Park

Exploring the wonders of Zion and Bryce Canyon National Parks reveals the magnificence of God's creation. Although these features are a result of the biblical Flood, this "recreated" landscape is immensely beautiful. When viewed from a biblical perspective, the parks have "God" written all over them. His wonderful handiwork can be seen in the grandeur of the vertical walls of Zion, the colorful hoodoos of Bryce Canyon, and the diverse and unique design of the plants and animals that inhabit these parks of splendor.

The parks are famous for their incredible, but very different, geologic features. Before these features were carved, over two miles of sedimentary rocks were deposited over the entire region. Subsequently, a massive erosional event removed about half of the material, leaving behind the stair step pattern called the "Grand Staircase." A series of five cliffs, totaling about 10,000 vertical feet of sedimentary layers, make up the "stairs," which extend over 60 miles across the west central portion of the Colorado Plateau. These sedimentary layers are capped by 2,000 feet of volcanic material on the northern-most stair (see diagram page 142).

Zion National Park symbolizes majestic beauty, size, and power. The park is filled with color, fantastic vertically walled cliffs, and alcoves that can truly inspire praise for God's creation. For this reason many of the features in the park are named after people or concepts from the Bible. Most of Zion National Park was carved out of the White Cliffs, which are made up of over 2,000 feet of sandstone, making them some of the tallest sedimentary cliffs in the world. Deep canyons dissect lofty mountains. Some, called "slot canyons," are hundreds of feet deep but only several feet wide.

Bryce Canyon National Park is a burnt orange and white tapestry of pinnacles called hoodoos within large amphitheaters. Many shapes can be imagined in the hoodoos, suggesting various animals or famous people. The park is carved in the Pink Cliffs, the highest "step" of the Grand Staircase. Unlike Zion, Bryce is mostly seen from above, along the rim of the canyon, but many trails allow you to experience the canyon from the bottom looking up at the amazing formations.

Please join us as we explore these awesome parks and view them from a biblical perspective.

Pronghorn antelope

Angels Landing, Zion

Inspiration Point, Bryce

SECTION ONE

About This True North Guide

Bryce

Zion

Gray fox

10

There are numerous books and guides written to help people understand Zion and Bryce Canyon National Parks. The vast majority of the interpretive literature available on these areas is based on evolutionary theory. The uniqueness of this *True North Guide* is in its perspective. *True North Guides* are written with three purposes in mind:

▶ *To provide an understanding of what the visitor actually sees from the major overlooks and stops along the way.*

▶ *To provide an overview of the Grand Staircase and associated parks from a creationist's perspective and compare it to the evolutionary interpretation.*

▶ *To show that what is observed in the Grand Staircase is best explained by the creation/Flood interpretation.*

WHERE TO BEGIN

Whether this is your first or 20th visit to the Grand Staircase region of south central Utah, this *True North Guide* is designed to help you get the most from your visit. It first provides some help in planning your trip and then an explanation of what you will see when you arrive. The large fold-outs for each of the overlooks and viewpoints in both parks present both site-specific information plus additional things to look for while visiting the Grand Staircase area. A summary of what you will find in sections two through eleven of this guide is provided in the following table.

2	**PERSPECTIVE** *page 14*	Presents an explanation of why Zion and Bryce Canyon National Parks are important from a biblical perspective.
3	**HOW TO SEE THE CANYONS** *page 18*	Provides the information on services and amenities in and around the parks, which will help you make the best use of your time.
4	**WHAT EVIDENCE WILL YOU SEE?** *page 26*	Summarizes some of the data seen at the parks that supports the biblical model. It will introduce you to ten evidences worthy of investigating as you consider the significance of the parks.
5 & 6	**EXAMINE EVERYTHING CAREFULLY** *Zion page 30* *Bryce page 74*	Contains a three-page foldout of all the major overlooks in both parks. Introductions include a suggested itinerary for visiting each park based on your available time. Each foldout offers a brief explanation of what is seen from the overlook with references to other sections for more in-depth information.
7	**UNDER-STANDING THE GRAND STAIRCASE** *page 134*	Provides some background for understanding not only what is seen here, but also the way in which it is interpreted. The section presents a foundation for the *Geology, Fossil,* and *Ecology* sections.
8	**GEOLOGY** *page 138*	Provides more detailed information on the geology observed at the parks than what is given on each overlook foldout.
9	**FOSSILS** *page 152*	Discusses general aspects of fossilization and offers more detailed information regarding fossils found in and around the parks.
10	**ECOLOGY** *page 156*	Provides more detailed information regarding the plants and animals inhabiting the parks than what is given on each overlook foldout.
11	**HISTORY** *page 162*	Offers a brief historical background of man's activities in and around the parks.

WHAT THIS GUIDE IS AND IS NOT

This *True North Guide* is not meant to be a technical publication, but presents the information in a way that a person with little or even no technical background can understand. Doing this requires the use of some terms that may be unfamiliar to the first-time visitor. Therefore, a glossary is included to provide definitions for many of these terms (see page 168).

Although the *True North Guide* was current when written, things do change and therefore it is not designed to replace the maps and information supplied by the National Park Service. When entering the park, we encourage you to obtain Zion National Park Service's *Zion Map & Guide* and the *Backcountry Planner*, and Bryce Canyon National Park Service's *The Hoodoo* newspaper. Further information may be obtained from the National Park Service's Internet sites at: www.nps.gov/zion for Zion National Park and www.nps.gov/brca for Bryce Canyon National Park.

> Zion: www.nps.gov/zion
>
> Bryce: www.nps.gov/brca

It is also not our intent to provide a complete list of lodging and amenities in and around the parks and other sites. Although we do list a few for your convenience, their listing does not imply any endorsement of their services or facilities. The National Park Service's internet site maintains a more comprehensive listing, along with most of the logistics necessary to visit the parks.

SAFETY TIPS

The Grand Staircase is an awesome place, but it can also be a dangerous place. In order to enjoy your time in the area, it is imperative you follow several safety rules.

▶ The most obvious is not getting too close to cliffs or venturing out past the guardrails on overlooks or along the rim. People fall and are injured or killed every year.

▶ Watch carefully for rattlesnakes and scorpions, especially down in Zion Canyon. Also, poison ivy grows in a few places in Zion.

▶ Carry more than enough water to keep yourself hydrated. (It is easy to underestimate and someone else may need some as well.) Most medical emergencies in this arid environment are related to dehydration.

▶ Use caution in all narrow canyons, which are subject to flash floods, often from unseen storms miles away.

▶ Do not remain on an isolated overlook, or under an isolated tree, during a lightning storm. Move to a more protected area until the storm passes.

▶ Don't throw rocks over the rim. There are people down there!

▶ In National Parks, pets must be on a leash and under physical control at all times. Pets are not allowed in public building or on trails, except the Pa'rus Trail in Zion National Park.

For emergency services, phone 911. For non-emergency health care, there is a clinic in Springdale just outside Zion National Park (435-772-3226) and both a hospital and clinic in Panguitch about 20 miles northwest of Bryce Canyon National Park (Hospital: 435-676-8811, Clinic: 435-676-2411).

STEWARDSHIP

The National Parks are places of beauty and wonder. With so many visitors each year, being good stewards of the earth, as the Bible says, is a necessity. By following National Park Service rules, we help preserve its beauty for others.

▶ Do not litter. In fact, should you find litter, please pick it up and deposit it where it belongs. Help keep our parks beautiful for the generations to come.

▶ Stay on the trails to protect fragile vegetation and reduce erosion.

▶ Don't deface the park by scratching your name on rocks, tree trunks, or park facilities. (Doing so carries a hefty fine.)

▶ Do not feed the animals. Not only is it dangerous for you, but "human food" can cause them sickness and malnutrition — and it is against the law.

If you take only pictures and leave only footprints, others will enjoy their stay more because you left no trace of your visit behind. As President Theodore Roosevelt said, "Leave it as it is."

The Subway, Zion

Perspective

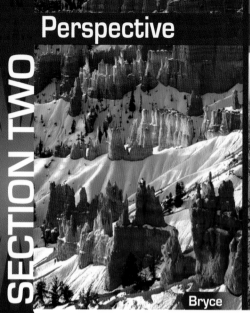

Bryce　**Zion**

Up until the 1800s, the dominant view in the Christian world of Europe and America was that God created the world in six normal 24-hour days about 4,000 BC.[1] According to this view, about 1,500 years later the earth was judged with a global, catastrophic flood during the time of Noah. This corresponds with the "young earth" hypothesis today that, based on the lineages laid out in the Bible, other historical documents, and scientific evidence, the initial creation occurred 6,000 to 10,000 years ago.

In the late 18th century, revisionist histories of the earth were written, most of which were naturalistic in philosophy, leaving God out of their ideas and eroding the foundation of Scripture. Naturalism is the philosophy that nature is all that exists — there is no supernatural. Scholars increasingly began to interpret earth history as a result of the slow, uniform, geological processes observed today and not as a result of the biblical Flood. This method of interpretation is known as uniformitarianism and is based on the concept that "the present is the key to the past." It theorizes that the processes operating today to modify the earth's surface are the same processes that operated in the geologic past.[1] It explains the origin and history of the earth by appealing only to time, chance, and the laws of nature working on matter.[2]

Uniformitarian thinking discounts major catastrophic processes, such as a global flood, in the development of geological formations. Thus, fossil beds and the sedimentary layers seen throughout the world have come to represent the millions of years of "geologic time" rather than representing rapid burial and sedimentation during Noah's Flood. God was denied, or at least His Word, the Bible, was left out of the equation in constructing this "new" history of the earth.

For simplification, we will normally use the term "evolutionary" when referring to uniformitarian thinking or processes. We understand this is not always technically the same, as some uniformitarians do not believe in biological evolution.

Although some have come to accept the Flood in Genesis chapters 6–9 as just a local flood, the Bible is clear it was a global event. This guide will go back to the historical roots of creation and the global flood. It will present evidence that a biblical understanding of the Grand Staircase is not in conflict with science, and that observable science supports the biblical Flood account in Genesis as being real history.

For example, consider the horizontal rock layers in the Grand Staircase (shown below). Notice the smooth surfaces between the layers, which are common

> *See to it that no one takes you captive through philosophy and empty deception, according to the tradition of men, according to the elementary principles of the world, rather than according to Christ.*
> Colossians 2:8

throughout the Colorado Plateau. Evolutionary geologists call these surfaces "unconformities" because they believe millions of years passed between the deposition of one layer and the next. In their view, these sedimentary layers were laid down when shallow seas advanced and retreated again and again across the continent to account for the sequence of layers found in the Grand Staircase and the mile of sediments found below them.

Layers of the Grand Staircase

If this were true, we would expect to see a considerable amount of erosion and uneven surfaces between the layers (see diagram page 146). Instead, the layers of the Grand Staircase look like pancakes stacked neatly on top of each other, and actually resemble what we see today within local flood deposits, only on a very large scale. This fact suggests little time elapsed between the deposition of one layer and the next, which is what we would expect to see if the layers were laid down rapidly in succession during a global flood event.

Creation geologists call these lines between the layers "contacts," which represent the boundaries between layers, but not periods of missing time (see page 146).

There are several other compelling geologic facts presented in section eight on the geology of the Grand Staircase (see page 138). They all demonstrate that true, observable science supports a straightforward reading of Scripture.

WHO BELIEVES?

It is often said, "There aren't any real scientists who believe in creation today." That is not true! There are many scientists today who operate from a creationist's point of view. Many of these hold M.S. and Ph.D. degrees from secular universities and see no conflict between science and God's Word. Likewise, many of the great founding fathers of science, like Isaac Newton, Louis Pasteur, Johannes Kepler, and Robert Boyle, found no conflict between their scientific research and their biblical worldview. It should also be noted that more and more evolutionary geologists are looking to catastrophic processes to explain the formation of the world's great rock units.

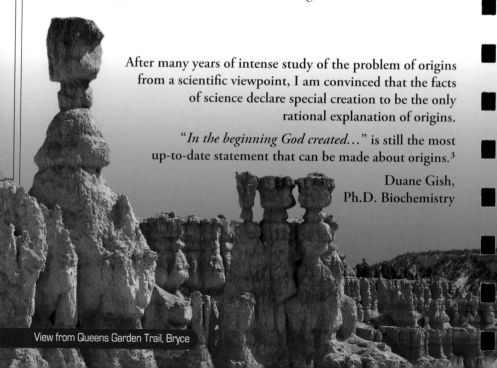

After many years of intense study of the problem of origins from a scientific viewpoint, I am convinced that the facts of science declare special creation to be the only rational explanation of origins.

"In the beginning God created…" is still the most up-to-date statement that can be made about origins.[3]

Duane Gish,
Ph.D. Biochemistry

View from Queens Garden Trail, Bryce

Zion Canyon from Zion-Mt Carmel Highway

WHY DO THEY BELIEVE?

It truly is a worldview issue. Most secular scientists do not believe supernatural events are part of the equation. For the Grand Staircase, this means the rock layers were laid down a particle at a time over hundreds of millions of years, and that Zion and Bryce Canyons were later carved slowly by gradual processes.

These theories tend to deny God's involvement and often His very existence. But if we look at this region from a biblical perspective, or worldview, we come to a very different conclusion based on what is observed.

The biblical record speaks of the Genesis Flood as a global flood in the time of Noah. In Genesis chapters 6–9, there are over thirty references to a worldwide flood. The context in the Bible is clearly that of a global, not local, catastrophe. If it were a local flood, the promise of the rainbow (Genesis 9), God's sign that He would never send another worldwide flood, makes no sense and would mean that God has broken His promise tens of thousands of times.

This catastrophic global flood was fed by water from two sources. One source was rain that fell for "forty days and forty nights" (Genesis 7:12). But the major water source was probably from beneath the surface of the earth as "the fountains of the great deep burst open" (Genesis 7:11). While there are differing opinions as to what "the fountains of the deep" were, most agree they were the source of a massive amount of water. There is sufficient water on the earth today that if the topography of the land were flattened to form a perfectly round sphere, it would be covered by water over one and a half miles deep. And if the pre-Flood mountains, which were likely destroyed during the Flood, were not as tall as we see today, there would be sufficient water to produce a global flood that covered "all the high mountains everywhere under the heavens" (Genesis 7:19).

So from a biblical perspective, the global flood of Noah's day provides a framework that explains the rock formations we see in the Grand Staircase and in the rest of the world as well. We contend that the scientific evidence found in the Grand Staircase supports and upholds the validity of the real history of the earth found in the biblical record.

How to See Zion and Bryce Canyons

Bryce   **Zion**

18

One can spend literally years exploring Zion and Bryce Canyon National Parks. There are overlooks to visit, trails to hike, wilderness backcountry to wander, and slot canyons to explore. Hikers can enjoy wild country and a great deal of solitude just a short distance from almost any road. These parks exhibit some of the most unusual erosional forms on the planet — truly a testimony to the power of water. In this section, we will examine how to best get around and experience the wonders of these amazing parks. The map on page 20 will provide you with an overview of the area.

The parks are open year round. Although wintertime activities are very popular in Bryce, some roads may be closed. A National Park pass or a weekly pass is required for entrance into the parks. Most areas are open to day use, but camping overnight along trails or in the backcountry requires a permit.

The visitor center is a good place to start your visit in either park. There you will find a wealth of information to help plan your visit based on your time and interests. Park rangers are available to answer questions. Also there are backcountry registration desks, orientation videos, book and gift stores, and interesting and helpful exhibits. The park newspapers are especially valuable to help plan your visit. You can access the newspapers online (see page 12 for addresses) or receive a copy as you enter the parks.

WORSHIP SERVICES

Several denominations hold worship services in the parks. See the park newspapers for times and locations.

WEATHER AND CLIMATE

The weather and climate vary considerably in Zion and Bryce Canyon National Parks — both seasonally and even daily. Visitors should always be prepared for extremes in the Grand Staircase area.

The table below shows the monthly average temperatures for each park. Remember that day-to-day weather can vary from the climatic average by a considerable amount. As you can see, temperatures in Zion, with its relatively low altitude, are hot in the summer, with maximum temperatures reaching over 110 degrees Fahrenheit. The corresponding temperatures at the higher altitude in Bryce Canyon are milder, with nighttime temperatures actually being quite cool during the summer. Winters are cold with potentially heavy snow in both parks, especially at higher altitudes. Spring is normally cool and dry with thunderstorms common from July into September, which can bring heavy rain and flash floods.

Month	ZION			BRYCE CANYON		
	High	Low	Precip	High	Low	Precip
January	52	29	1.6	39	9	1.7
February	57	31	1.6	41	13	1.4
March	63	36	1.7	46	17	1.4
April	73	43	1.3	56	25	1.2
May	83	52	0.7	66	31	0.8
June	93	60	0.6	76	38	0.6
July	100	68	0.8	83	47	1.4
August	97	66	1.6	80	45	2.2
September	91	60	0.8	74	37	1.4
October	78	49	1.0	63	29	1.4
November	63	37	1.2	51	19	1.2
December	53	30	1.5	42	11	1.6
Annual Average	75	47	14.4	60	37	16.3

Average temperatures in Fahrenheit, precipitation in inches

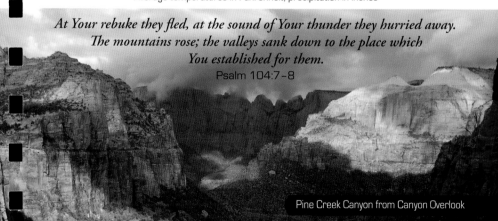

At Your rebuke they fled, at the sound of Your thunder they hurried away.
The mountains rose; the valleys sank down to the place which
You established for them.
Psalm 104:7–8

Pine Creek Canyon from Canyon Overlook

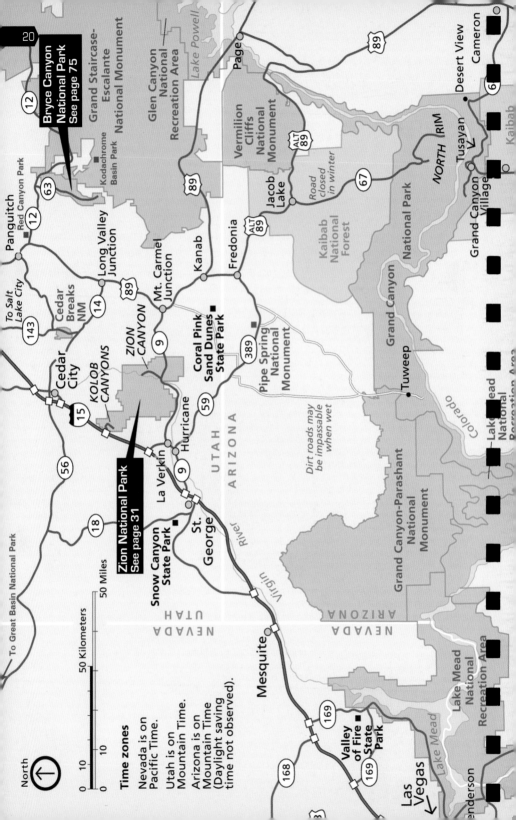

Bryce Canyon National Park See page 75

Grand Staircase-Escalante National Monument

Glen Canyon National Recreation Area

Lake Powell

Page

89

89

Desert View

Cameron

Kaibab

Vermilion Cliffs National Monument

ALT 89

Road closed in winter

NORTH RIM

Tusayan

Grand Canyon Village

12

Jacob Lake

67

Kaibab National Forest

Grand Canyon National Park

63

Kodachrome Basin Park

Panguitch

Red Canyon Park

12

89

Fredonia

ALT 89

Kanab

Mt. Carmel Junction

Long Valley Junction

Cedar Breaks NM

14

89

Grand Canyon

Colorado

143

To Salt Lake City

ZION CANYON

9

Coral Pink Sand Dunes State Park

389

Pipe Spring National Monument

Tuweep

Lake Mead National Recreation Area

Cedar City

KOLOB CANYONS

59

Dirt roads may be impassable when wet

15

Hurricane

UTAH

ARIZONA

56

La Verkin

9

18

Zion National Park See page 31

Snow Canyon State Park

St. George

Virgin River

Grand Canyon-Parashant National Monument

UTAH

NEVADA

ARIZONA

NEVADA

Time zones

Nevada is on Pacific Time.

Utah is on Mountain Time.

Arizona is on Mountain Time (Daylight saving time not observed).

To Great Basin National Park

Mesquite

Virgin

Lake Mead National Recreation Area

Lake Mead

169

169

Valley of Fire State Park

168

169

Las Vegas

Henderson

North

50 Miles

50 Kilometers

50

10

10

0

0

ZION NATIONAL PARK

Zion National Park has some of the tallest sandstone cliffs in the world. The majesty of clear blue skies and colorful cliffs, all accented by an array of green trees, is likely to stop Zion visitors in their tracks. Zion is a Hebrew word that not only is a synonym for Jerusalem, but also for the Promised Land to come. Zion was interpreted to mean a "place of safety or refuge" by Mormon pioneers who gave the name to the canyon in 1860.

SPRINGDALE

Just outside the park entrance is the small town of Springdale. The town offers many amenities, including motels, campgrounds, several charming art galleries, gift shops, rock shops, an IMAX theater, and a variety of dining choices. Springdale also hosts festivals and fairs, including live theater, and has an outdoor amphitheater for concerts. Check brochures available in most restaurants and motels for details. There are also "adventure sports" businesses with organized tours, rentals, and sales to help visitors fully enjoy activities both in and outside the park. Services include equipment rental to hike The Narrows in comfort, river tubing, mountain biking, canyoneering, and backpacking. A two-mile, two-hour tubing trip down the Virgin River through Springdale is a refreshing way to spend a hot summer afternoon.

ZION CANYON VISITOR CENTER

To orient your visit, the Zion Canyon Visitor Center, as well as the Human History Museum, are good places to start. Use the exhibits to make plans based on your time and interests (see suggested itineraries on page 32). Park personnel and rangers can advise you of activities within the park such as guided hikes, daytime talks, and evening programs. Also, summertime children's programs are conducted twice daily at the Nature Center just north of the visitor center.

The architecture of the visitor center is an interesting study in energy conservation. It uses passive down-draft evaporative cooling towers, day lighting, natural ventilation, and solar heat collecting walls for temperature control, both winter and summer. Photovoltaic panels provide the majority of the electricity needed by the building. Outside, gravity-flow irrigation ditches provide most of the water needed for native trees and landscaping.

ZION HUMAN HISTORY MUSEUM

The museum is the first shuttle bus stop from the visitor center, but you can drive to it as well. The 20-minute orientation film starts on the hour and half hour. There are interesting exhibits concerning Native Americans, Mormon pioneers, local plants and animals, and the use of water in this arid land. Please see page 34 for more information.

WAYS TO SEE ZION NATIONAL PARK

Here are a few of the many ways to see and enjoy Zion. For suggested itineraries based on your available time, read *Examine Everything Carefully– Zion National Park* section (see page 30).

Shuttle Bus: From March through October, private vehicles are not allowed to drive through Zion Canyon; instead, the park service provides a free shuttle service with natural gas powered buses. A shuttle departs from each stop about every seven minutes and you can get on and off as often as you wish. A separate shuttle loop runs through the town of Springdale and drops visitors near the park's main gate. The two shuttle loops are connected by a footbridge between the visitor center and the Zion Canyon IMAX Theater.

Biking: Bicycles are permitted only on established roads and the Pa'rus Trail. Ride defensively and considerately. Park shuttle buses will not pass bicyclists; you must pull over and stop to let them pass. See the *Zion Map and Guide* for further information.

Hiking: There are many hikes to match your interest depending on your experience, physical condition, available equipment, and the weather. There are popular short hikes of less than a quarter of a mile to Weeping Rock and The Court of the Patriarchs. Then there are moderate hikes of one to four miles, such as the trail to Hidden Canyon, up the Virgin River to The Narrows, and the hike to Emerald Pools. Many longer hikes such as Angels Landing (see page 46) and The Subway are available.

Horseback Riding: Guided rides are available March through October. Inquire at the visitor center or at the Zion Lodge. Reservations are advised.

BRYCE CANYON NATIONAL PARK

At Bryce Canyon, erosion has shaped amphitheaters filled with towering hoodoos alive with abundant colors and tints too numerous to name. Bryce's hoodoos are world-renowned and among the most photographed features in the Southwest.

RUBY'S INN

Ruby's Inn and Campground is a hub of services and activities for visitors. Just a mile north of the park, you will find motel rooms, large RV parking with hookups, tent sites, tepee and cabin rentals, a general store, an art gallery and gift shop, two swimming pools, and a host of other amenities. Across the street are rodeo grounds, unique shops in a western setting, and another motel.

BRYCE CANYON VISITOR CENTER

Just inside the park, across from the fee station, is Bryce Canyon Visitor Center. The center is a great place to talk to rangers and other park staff who can help you plan your visit. In addition to the bookstore, there is a 20-minute orientation film that plays on the hour and half hour. There are interesting exhibits including a large relief map that will orient you with respect to Zion and the Grand Canyon. Like other National Parks, Bryce has many interesting ranger-led activities available from mid-May through August.

STAR GAZING

Bryce Canyon rangers and volunteers offer astronomy programs during the summer. Natural darkness is becoming increasingly rare, but Bryce has some of the darkest, starriest night skies in the nation, providing excellent star gazing opportunities. The talks start with an indoor multimedia presentation, sometimes requiring advance free tickets. Following the indoor presentation, there is outdoor telescope viewing of the stars (weather permitting). Check at the visitor center or on the website for program schedules.

WAYS TO SEE BRYCE CANYON

There are many ways to see and enjoy Bryce. Here are just a few of them. Suggestions of itineraries are in the *Examine Everything Carefully–Bryce National Park* section (see page 74).

Shuttle Bus: Though you are allowed to drive to Bryce's many overlooks, parking and traffic can be a problem during the busy season. A free shuttle bus is available that goes from Ruby's Inn to Bryce Point, stopping at all the major overlooks in between. The buses run from late May to September. Detailed information is available in the park newspaper or at the visitor center.

Biking: Bicycles are restricted to paved roads in the park, and most of the overlooks can be accessed. Caution is necessary, as the roads are winding with many blind curves. There are also excellent mountain biking opportunities just outside of the park on single and double track trails and forest roads.

Hiking: There are a variety of hikes in Bryce. Short hikes include the Bristlecone Loop Trail at the far southern end of the park and the Rim Trail between Bryce and Fairyland Points, which traverses through the hoodoos. Several trails lead into the canyon and therefore are a bit more strenuous. They include the Navajo Loop Trail, the Queens Garden/Navajo Combination Loop Trail, the Peekaboo Loop Trail, and the Fairyland Loop Trail.

Horse Rides and Air Tours: Two- and four-hour horseback rides into Bryce Canyon are available through the Bryce Canyon Lodge and reservations are suggested (see www.canyonrides.com.) Outside the park at Ruby's Inn, arrangements can be made for horseback rides and air tours.

OTHER THINGS TO DO IN THE AREA

There are a number of interesting things to do around the Grand Staircase area (see map on page 20). There are thousands of miles of hiking trails on public lands around Zion and Bryce Canyons. Many of the trails and roads are open to off-road enthusiasts. More information is available in the brochure stands located in many of the shops and motels in the area or online at their various web sites. Following is a short list of some of the nearby parks and attractions you may wish to visit.

Cedar Breaks National Monument: Carved in the Pink Cliffs at the top of the Grand Staircase, Cedar Breaks is north of Zion off Utah Route 14. At an elevation of 10,460 feet, it is like a miniature Bryce Canyon with a large sculptured amphitheater 2,000 feet deep and more than three miles across.

Bristlecone pine trees (see page 160) are relatively common at Cedar Breaks. They and many wildflowers are best seen on short hikes on the southern rim to Spectra and Rampart Points.

Kolob Canyons: Most people miss this littleknown northwest section of Zion National Park. It displays tall vertical cliffs with hanging valleys. Plan on at least four hours round trip for the drive west then north from Springdale. From Kolob Canyons, there is a strenuous 14-mile round trip hike to Kolob Arch, the second largest arch in the world (see page 147).

Red Canyon National Forest Park: Ten miles west of the turnoff to Bryce Canyon on Utah Route 12, Red Canyon is a fascinating area set aside and administered by the Dixie National Forest. Often called "Little Bryce," the area has bright orange-red hoodoos, arches, and cliffs amidst towering ponderosa pines.

Kodachrome Basin State Park: This is a small Utah State Park located 20 miles southeast of Bryce Canyon National Park, off Utah Route 12. The park is known for its colorful monolithic spires, some reaching 170 feet high. The origin of sandstone spires, more correctly called liquefaction plumes, are an enigma for evolutionary geologists to ponder.

Great Basin National Park: Established in 1986, Great Basin National Park is a relatively new addition to the park system. It is a mountain in a sea of sagebrush. The park is a real treasure for those interested in the natural history of the Great Basin region. The park includes Lehman caves, which boast a wide assortment of stalactite and stalagmite formations.

Coral Pink Sand Dunes State Park: These salmon-colored shifting sand dunes — the color of which is not found at any other place in the world — is a massive playground for hiking, photography, and off-road vehicles. At 6,000 feet in elevation, the park encompasses more than 3,000 acres of sand dunes and is open year-round.

Mossy Cave: This arching cave is a short hike at the north end of Bryce Canyon National Park. The trail passes a small waterfall, a fun place to cool off in the summer heat. It can be accessed by driving out of Bryce and east on Highway 12 about four miles toward the town of Tropic. The trailhead and a small parking area are on the right side of the highway.

Grand Staircase – Escalante National Monument: A vast area located southeast of Bryce Canyon National Park offers a variety of scenic and recreational opportunities.

Escalante Petrified Forest State Reserve: Located in Escalante State Park, the reserve displays colorful deposits of petrified wood and mineralized dinosaur bones. It also has a popular bird watching location, a reservoir for recreation, 1,000-year-old petroglyphs, and Fremont Indian relics.

Le Fevre Overlook of Grand Staircase: Located on Highway 89A 40 miles southeast of Kanab, Utah, the overlook provides a panoramic view of many of the colored rock layers of the Grand Staircase.

St. George Dinosaur Discovery Site at Johnson Farm: See a large enclosed fossil display in St. George, Utah, including some extraordinary and very rare dinosaur tracks.

Mountain lions live in and around both parks.

What Evidence Will You See?

Bryce

Zion

God's handiwork around us displays His glory. As you visit these canyons, we hope you will examine carefully what you see and hear, and we invite you to do so from a biblical worldview. As Paul stated, we should *"examine everything carefully; hold fast to that which is good"* (1 Thessalonians 5:21).

The following table is a comparison of two different worldviews you might expect to find in viewing the world around you. As you review the table and others like it throughout the text, consider what the physical evidence actually shows and how it might be interpreted.

WORLDVIEW EXPECTATION TABLE		
If the earth were the result of processes acting over a long period of time, versus the result of a supernatural creation and a worldwide flood, the following would be expected:	**EVOLUTION** ▼	**CREATION** ▼
	long period of time	recent supernatural creation
The worldwide geologic record would:	show abundant evidence for slow, gradual processes	show evidence of rapid catastrophic processes
The fossil record would:	reveal abundant transitional life forms	lack transitional life forms
Biological systems would:	become more complex with time	be created complex and fully functional initially
Geologic features would be:	local in nature	large-scaled in nature
Intelligence in the animal kingdom would be:	learned	inherited

KEY POINTS

This guide will showcase the following key points that support the creation/flood model and refute millions of years and evolution. They are summarized here, but explained in more detail in later sections.

Extent of Sedimentary Layers: The horizontal layers of the Grand Staircase can be traced hundreds of miles, unlike the depositing of sediments seen today. (See page 145.)

Boundaries between Layers: Were the layers deposited one on top of another separately over hundreds of millions of years or as one continuous vertical sequence? The evidence suggests the latter, because the horizontal lines (contacts) between the individual sedimentary rock layers show little or no sign of erosion between them. (See page 146.)

> As you visit these canyons, we hope you will examine carefully what you see and hear, and we invite you to do so from a biblical perspective.

Narrow, Vertically Walled Canyons: Such canyons are thought to be millions of years old, but the evidence suggests they are young since vertically walled canyons become more V-shaped with time. The canyons of Zion look just as expected if cut rapidly by the channelized flow of receding water late in the Flood, which occurred about 4,500 years ago. (See page 144.)

Lack of Rockfall: There is a significant lack of rockfall debris on the slopes at the base of the cliffs throughout Zion National Park. A recent and catastrophic carving of the canyon is the key to this lack of rockfall. (See page 144.)

Rock Arches: Freestanding arches need to have formed quickly and not long ago or else erosion would have caused their collapse. These types of arches, different from the windows commonly seen in Bryce Canyon, are evidence of massive erosion during the Flood. (See page 147.)

Lack of Transitional Fossils: Just like everywhere else on the earth, there is a lack of transitional fossils between distinct categories of plants and animals. (See page 153.)

Zion's historic Crawford Arch, also known as Bridge Mountain Arch, is only a few feet wide and over 150 feet long. The arch may be seen high on the east wall of Zion, across from the Human History Museum.

Fossil Record Support of the Creation Model: The conditions necessary to form a fossil are rare today. Besides rapid burial, certain minerals must be absorbed quickly into the bone, wood, or shell to preserve the fossil, but groundwater is practically always too low in these minerals. Thus, billions of fossils in the rocks attest to mass burial and rapid absorption of abundant minerals under high pressure during a catastrophic event. (See page 152.)

Fossil Communities Uncommon: Fossils are not found in "life assemblages," (i.e., with fossils of creatures that normally live together). Instead, fossils are sorted as would be expected by a watery catastrophe. (See page 154.)

Amazing Adaptations of Desert Plants: Desert plants have ingenious adaptations to survive in a hot, dry environment. They have diverse mechanisms to stay cool, find and store water, and prevent water loss. It does not make sense that such fine-tuned adaptations could evolve. They must have been created with all these marvelous adaptations already in place. (See page 158.)

Yucca Plant and Moth: Certain species of plants and animals cannot live without each other. For example, the banana yucca and the yucca moth depend upon each other for survival. This co-dependency with its many interdependent parts must have been created fully functioning. (See page 160.)

One of the many walls of hoodoos seen in Bryce Canyon

WHAT IS THE BOTTOM LINE?

So what is the bottom line, the true message of these amazing parks? As you examine the geology of the Grand Staircase area, there are really only two choices: it is either a monument to time, or a monument to the Flood. The evidence suggests the latter.

As a monument to the Flood, it is the result, not of God's initial creation, but of His judgment of a world marred by sin. The Grand Staircase stands before us today as evidence of the time when *"...the wickedness of man was great in the earth...and God said...behold, I am about to destroy them with the earth...behold I, even I am bringing the flood of water upon the earth, to destroy all flesh in which is the breath of life..."* (Genesis 6:5, 13, 17).

Therefore, as you view the splendor seen in Zion and Bryce Canyon National Parks and the rest of the world, remember that the rocks are a "re-creation" of the earth's surface after a global judgment by water.

Although we see amazing beauty, it is likely but a glimpse of the beauty His original creation displayed. Everything around us has been marred by sin that came about because of a real historical rebellion, by a real historical man, Adam, followed by the rebellion that led to man's destruction by the Flood.

Because of this, God sent His real, historical Son, the Lord Jesus Christ, into the real world of space and time, to bear our sins on the Cross. That is a powerful message to consider as you visit Zion and Bryce Canyon National Parks.

Great is the Lord, and greatly to be praised.
Psalms 48:1

SUGGESTED ITINERARIES

The exhibits at the visitor center will help you plan your visit based on your interests and available time. Some recommendations are briefly summarized below.

Less than three hours: Take the park shuttle on the Zion Canyon Scenic Drive. It will allow you to see some of the park's most beautiful sights in a short amount of time. A round trip, without getting off, takes a minimum of 90 minutes. Popular stops include the lodge, a short hike to Weeping Rock (see page 50), and the Court of the Patriarchs (see page 38). The Zion-Mt. Carmel Highway includes a mile-long tunnel as the road winds through beautifully swirled and cross-bedded sandstones and past Checkerboard Mesa (see page 66), just inside the east entrance. The one-mile round trip hike to Canyon Overlook (see page 62) provides a glorious view of Zion National Park.

Half a day: Begin with the suggestions above, then consider spending some time at the visitor center and the Human History Museum. Add a short hike up Riverside Walk at the end of Zion Canyon Scenic Drive (see page 59).

Full day or more: In addition to the suggestions above, just relax and soak in the grandeur of the canyon at Zion Lodge and take a hike to Emerald Pools or Hidden Canyon. Or get up early and take one of the more adventurous hikes — a five-mile round trip hike to the top of Angels Landing. This hike is not for those fearful of heights, as the last half-mile traverses a steep, narrow ridge with a chain railing. Another possibility is a drive to Lava Point in the northwest part of the park. The drive starts west of Springdale on Utah Route 9 and winds through the high country, providing spectacular views of the entire region (see page 70).

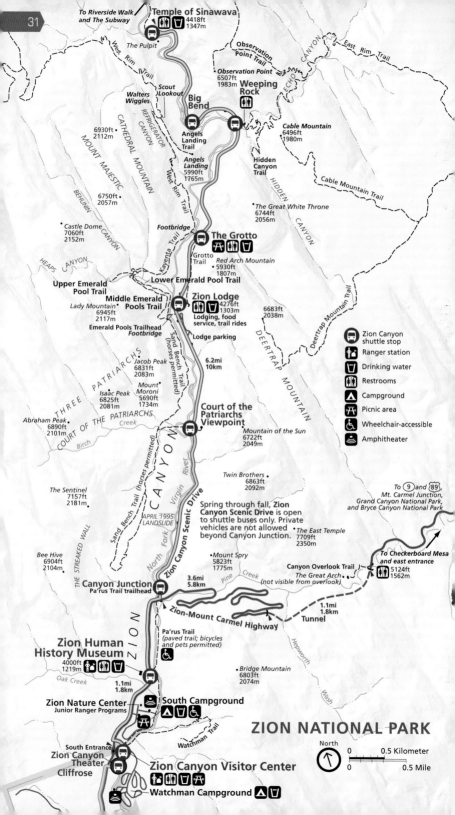

To Riverside Walk and The Subway

Temple of Sinawava
4418ft
1347m

The Pulpit

Observation Point Trail

Observation Point
6507ft
1983m

Weeping Rock

Scout Lookout

Walters Wiggles

Big Bend

6930ft
2112m

Angels Landing Trail

Cable Mountain
6496ft
1980m

Angels Landing
5990ft
1765m

Hidden Canyon Trail

MOUNT MAJESTIC

CATHEDRAL MOUNTAIN

REFRIGERATOR CANYON

6750ft
2057m

The Great White Throne
6744ft
2056m

BEHUNIN CANYON

Castle Dome
7060ft
2152m

Footbridge

The Grotto

Grotto Trail

Red Arch Mountain
5930ft
1807m

HEAPS CANYON

KAYENTA TRAIL

Lower Emerald Pool Trail

Upper Emerald Pool Trail

Middle Emerald Pools Trail

6683ft
2038m

Lady Mountain
6945ft
2117m

Zion Lodge
4276ft
1303m

Lodging, food service, trail rides

Emerald Pools Trailhead *Footbridge*

Lodge parking

6.2mi
10km

Jacob Peak
6831ft
2083m

Mount Moroni
5690ft
1734m

THREE PATRIARCHS

Isaac Peak
6825ft
2081m

Court of the Patriarchs Viewpoint

DEERTRAP MOUNTAIN

Abraham Peak
6890ft
2101m

COURT OF THE PATRIARCHS

Birch Creek

Mountain of the Sun
6722ft
2049m

Twin Brothers
6863ft
2092m

The Sentinel
7157ft
2181m

APRIL 1995 LANDSLIDE

Spring through fall, **Zion Canyon Scenic Drive** is open to shuttle buses only. Private vehicles are not allowed beyond Canyon Junction.

The East Temple
7709ft
2350m

To ⑨ and ⑧⑨
Mt. Carmel Junction, Grand Canyon National Park, and Bryce Canyon National Park

Bee Hive
6904ft
2104m

Mount Spry
5823ft
1775m

Canyon Overlook Trail
The Great Arch
(not visible from overlook)

To Checkerboard Mesa and east entrance
5124ft
1562m

THE STREAKED WALL

Canyon Junction
Pa'rus Trail trailhead

3.6mi
5.8km

Zion-Mount Carmel Highway

1.1mi
1.8km
Tunnel

Pine Creek

North Fork Virgin River

Sand Bench Trail (horses permitted)

Zion Canyon Scenic Drive

Zion Human History Museum
4000ft
1219m

Pa'rus Trail
(paved trail; bicycles and pets permitted)

Bridge Mountain
6803ft
2074m

Oak Creek

1.1mi
1.8km

Zion Nature Center
Junior Ranger Programs

South Campground

Hepworth Wash

ZION CANYON

Watchman Trail

Zion Canyon Theater
Cliffrose

South Entrance

Zion Canyon Visitor Center

Watchman Campground

ZION NATIONAL PARK

Legend:

- 🚌 Zion Canyon shuttle stop
- Ranger station
- 🚰 Drinking water
- 🚻 Restrooms
- ⛺ Campground
- ⛱ Picnic area
- ♿ Wheelchair-accessible
- Amphitheater

North

0   0.5 Kilometer
0   0.5 Mile

Examine Everything Carefully

SECTION FIVE

Zion National Park

This section will allow you to examine the splendor of Zion National Park in more detail as you stop at the overlooks and points of interest. There is a three-page layout for each of the major overlooks and bus stops in the park. These will provide you with an understanding of what is seen, along with references to other sections for more detailed information. A variety of subjects are discussed, including the geology, ecology, hiking, history, and archeology. This park is at a lower altitude than most of Bryce Canyon National Park, so the trees, animals, and weather are generally different.

The majority of visitors enter the park through Springdale. From them the best place to start your tour is at the visitor center (see page 21). Then you can drive or take the shuttle to the first stop, the Zion Human History Museum. The other overlooks in Zion Canyon are closed to private vehicles from March to October, except if you have reservations at Zion Lodge. The National Park Service runs a free shuttle service during those months to accommodate the approximately three million visitors.

To get an overall view of Zion National Park and its geological and geographic context, a drive to Lava Point should be considered. You will see canyons carved over 3,000 feet deep in the Kolob Plateau, leaving a series of tall, generally flat-topped mesas. You will also see that Zion National Park is part of the Grand Staircase, a series of five cliffs or steps caused by the erosion of over 10,000 feet of sedimentary rocks to the south. Zion is carved on the third step from the bottom, called the White Cliffs, named for the white Navajo Sandstone.

Zion National Park from Canyon Overlook

If you enter by the east entrance, your first stop can be Checkerboard Mesa (see page 66). The park signs at Checkerboard Mesa describe the main rock unit in the park, the Navajo Sandstone, easily identified here by its colorful swirled and cross-bedded layers. As you proceed toward the tunnel, watch for desert bighorn sheep; Canyon Overlook is just before you reach the east entrance of the tunnel, parking is available for a half-mile hike to a spectacular overlook of Zion Canyon. California condors are often seen soaring along the cliffs in this area.

Services within the park are located at the visitor center and Zion Lodge. The only food available in the park is at the lodge. The Park Service campgrounds are located near the visitor center. Just outside the west entrance is the town of Springdale, catering to tourists with restaurants, lodging, and shopping. Parking can be difficult at the visitor center, so there is also the option of parking in Springdale and taking the free town shuttle to the visitor center where you can catch the shuttle bus into Zion Canyon. For further ideas on how to plan your visit, read section three of this guide, *How to See Zion and Bryce Canyons*, on page 18.

As you view Zion from the overlooks, you will notice many of the same features are seen from most locations throughout the park, but if you visit all the overlooks and bus stops, you will see them from several perspectives. Once you understand the overall formation of Zion Canyon, you will be able to recognize these common features. Each three-page layout will suggest interesting things to observe and some questions to consider.

We hope you will enjoy this different perspective of Zion National Park.

C. Oak Creek Canyon,
site of late 1800's diversion ditch
and dinosaur tracks

D. Altar of Sacrifice,
7,505 feet above sea level, with
iron oxide stains streaming down
from the top

MUSEUM

seen across the
the museum

Mormons to irrigate orchards and crops in the late 1800s. Flats like this have been good locations not only for the pioneers, but also for the Paiutes and the Anasazi who came before them (see page 162). With water and a long growing season, Zion Canyon provided fertile ground for needed crops.

FOSSILS: In the canyon behind the museum, as well as near the trail to The Subway, dinosaur tracks have been discovered (but are very difficult to find as they are not well documented). Dinosaur tracks occur by the millions all over the world. Trackways of several prints from the same dinosaur are nearly always headed in a straight line, indicating that they were likely fleeing something. Dinosaur tracks can be explained by the creatures fleeing the Flood on sediments briefly exposed due to a local drop in sea level[5] (see page 155).

ECOLOGY: Cacti have the ability to expand their stems to store water when it is available. Also, cacti, agaves, and yuccas amazingly can grow "rain roots" within hours of the soil getting soaked. These fine-haired shallow roots take advantage of thunderstorm water but die back in times of drought.

Dinosaur tracks

A. West Temple,
7,810 feet above sea level and the highest point in southeastern Zion

B. Vertical walls,
indicating a rapid erosional process

Crawford Arch, valley from

HUMAN HISTORY

GEOLOGY: From the backside of the museum, two miles to the west, Zion's West Temple (A) rises majestically 3,800 feet from the valley floor. This is the greatest single vertical rise in the park. Notice the Altar of Sacrifice (D), named for the blood-red iron stains streaking down its surface.

To the west-northwest, the peaks of the Towers of the Virgin frame the skyline. Interestingly, this skyline is only a ridge with similar, mostly dry, vertically walled canyons on the other side. Many side canyons in Zion have small headwaters and are often dry. Such narrow, vertical-walled canyons are evidence of young, rapidly incised canyons,[4] eroded by lots of water (see page 144).

In front of the museum, a park sign points out an amazing free-standing arch (pictured above) seen high on Bridge Mountain across the canyon. This is the Crawford Arch. Arches like these are signs of a young landscape (see page 147).

HISTORY: Up Oak Creek Canyon (C), below the Towers of the Virgin, are the remains of a pioneer diversion ditch built by the

HUMAN HISTORY MUSEUM

The Human History Museum is the first shuttle stop after the visitor center. At this stop, you can see beautiful Zion Canyon photography and learn about the history of the park in a 20-minute orientation film. Exhibits include information on the ancestral Puebloans (Anasazi), the Southern Paiutes who came after them, and the Mormon pioneers. Information on park plants, animals, and ecology is displayed. The landscape outside of the museum is practically a museum in itself.

FAST FACTS
> Dinosaur tracks are found nearby.
> The skyline to the west is only a thin divide.
> Narrow, vertically-walled valleys indicate rapid erosion by a lot of water.
> The datura flower is pollinated at night by the sphinx moth.

The Zion Human History Museum has many interesting artifacts from the Native Americans who first settled this region.

Question: Where is the tallest single vertical rise in the park?

E. Landslide debris,
probably from post-flood slides

F. Beehives,
6,904 feet above sea level,
top of the Navajo Sandstone

ELEVATION 3,993 feet

Sacred datura, also known as jimson weed or moonflower, is found growing throughout the canyon. This poisonous plant has fragrant, white to violet, trumpet-shaped flowers, which open at night and wither by mid-morning. They are pollinated at night by the sphinx moth and other insects.

Used ceremonially and medicinally by Native Americans, even the "right" dose of a datura with stronger than average poison can cause death. Just touching this beautiful plant can cause a nasty rash. So, be careful!

Sphinx moth

A. Abraham Peak, 6,890 feet above sea level, named after the patriarch Abraham in the Bible

B. Isaac Peak, 6,825 feet above sea level, named after Abraham's son and topped by pine trees

TRIARCHS

FOSSILS: Bordering the trail leading to the viewpoint are some interesting sandstone blocks. Some have fossilized tracks (left) on them. One wonders if they might have been "escape tunnels" of small creatures trying to escape the Flood.

ECOLOGY: Wild turkeys were reintroduced to Zion in the 1950s and are seen on the valley floor. Females are often seen in small flocks, while the males are more solitary. Anthropologists have discovered that turkeys were penned and raised by Anasazi Indians.

Male wild turkey

Runoff from hanging valley.

GEOLOGY: Between the Human History Museum and Court of the Patriarchs, the road passes through a landslide area. The landslide, which likely occured well after the receding of the Flood, dammed the river, creating Sentinel Lake. Before the lake broke through, sediments built up in the lake bottom, creating the flat valley floor. The silty soils stretch past The Grotto. As recently as 1995, the edge of the old landslide slid again, diverting the river and washing out part of the road.

Looking across the valley are the three patriarchs, Abraham, Isaac, and Jacob, named by Reverend Frederick Vining Fisher in 1916 after the biblical patriarchs from the book of Genesis. The Patriarchs rim the north side of Birch Creek Canyon, which like many canyons in the area are steep-walled — a sign of a youthful canyon. Consider what might have formed this side canyon. Like many other side canyons, it was not carved by the Virgin River and probably not by streams flowing from above.

The rims of Zion Canyon are scalloped by shallow tributary valleys between mountain peaks. These upper valleys, whose floors are high above the main valley, are called "hanging valleys" (C). Creationists believe these upper valleys were eroded near the end of the Flood,

COURT OF THE PA

during the uplift of the Colorado Plateau. As the waters began to channelize, downcutting was accelerated in the main canyons and diminished in the upper canyons. This left the lesser upper valleys hanging on the main canyon walls. Waterfalls cascade from seemingly nowhere, as they gush from the mouths of these hanging valleys during and after rainstorms (above).

Behind you on the east wall is a hanging valley with the black stains of the iron/manganese oxides from the seasonal runoff water marking the rock surfaces. The white streaking seen across the canyon is left by calcium and magnesium carbonates in water as it evaporates on the way down.

Fossilized worm tracks

To view the Court of the Patriarchs, take the short trail up the hill from the bus stop. The three peaks, named by Reverend Fisher, are formed in the Navajo Sandstone, which is composed of sets of cross-beds separated by bounding surfaces (see page 149). Hanging valleys are seen between the peaks, especially to the left of Abraham Peak. As you continue into the valley, keep an eye out for mule deer, wild turkeys, foxes, or coyotes. Because of its relatively low altitude, cacti can also be seen in this area.

FAST FACTS

> Hanging valleys could form during the receding of Noah's Flood.
> Turkeys were penned and raised by the Anasazi Indians.
> In 1995, a large landslide near here caused the Virgin River to wash out a section of the road.
> White-tailed antelope squirrels have interesting temperature-coping mechanisms.

Prickly pear flowers have an amazing mechanism to aid pollination. They respond to touch. If you poke your finger into a flower, the stamens bend inward, dusting your finger with pollen. This response ensures that a bee will be coated with pollen before flying off to the next flower.

Engelmann's prickly pear cactus and flower

Question: Why is the valley floor flat between here and Angels Landing?

C. Hanging valley,
carved by Flood water
draining into Zion Canyon

D. Jacob Peak,
6,831 feet above sea level, named
after Abraham's grandson

ELEVATION 4,195 feet

The white-tailed antelope squirrel's body
temperature can rise to 109 degrees without
ill effects. They have an interesting temperature-
coping strategy. They drool large amounts of saliva and
rub it over their heads with their forepaws to let the
evaporation cool them. This uses the same evaporative
principle as swamp coolers used in homes in the
southwest and the cooling towers at the visitor center
(see page 21).

But ask the animals, and they will teach you...
        Job 12:7

A. Snack bar,
with a grill, ice cream, frozen
yogurt, and an outdoor patio

B. Main lobby,
with a gift shop and
restaurant upstairs

Dragonflies are one of the most remarkable creations in the insect world. They have a completely different mechanism for flying than any other of the 800,000 insects. Most insects' flight is controlled by muscles in the thorax area, which push the wing up when tightened and down when relaxed. In contrast, the dragonfly has muscles connected directly to the wing joints by tendons, giving them eight modes of flight, including flying backwards. Pioneer designer Igor Sikorsky got the idea for the development of the helicopter from his observations of the dragonfly.[6]

Red skimmer dragonfly

Emerald Pools

HISTORY: Mormon pioneer Isaac Behunin built a cabin and grew crops near the present location of Zion Lodge in the 1860s. Behunin is credited with first giving Zion Canyon the name "Zion," which among other meanings refers to the Promised Land to come in which God dwells among His people.

GEOLOGY: Take a minute and look at the surrounding cliffs. Over 2,000 feet thick here in Zion, the Navajo Sandstone is one of the tallest solid sandstone formations in the world. Where not eroded away, this layer actually spreads out over 130,000 square miles of the Colorado Plateau. Where did all this sand come from in the first place? There are major scientific problems with the prevailing story that says the Navajo Sandstone developed in a desert dune environment (see page 142).

HORSEBACK RIDING: The horse corral is across the street just south of the bridge. You can enjoy a one- or four-hour ride through the canyon with a qualified guide.

ECOLOGY: A huge Freemont cottonwood tree (D), seven feet in diameter at the trunk, is the centerpiece of the lawn area in front of the lodge where people find comfort in its shade. On a hot day, a tree of this size transports over a thousand gallons of water to its leaves. It takes a lot of energy for a machine to pump water a hundred feet above the ground, but large trees are able to lift water up by three remarkable mechanisms: capillary action, osmosis, and vacuum pressure. The engineering excellence of this system is a witness against the idea that it could have developed by chance.

ZION LODGE

In desert country, the presence of cottonwood trees says "water is here," even if you have to dig for it. The trees also provide welcome shade and homes for many birds that prefer to nest in big trees. Cottonwood leaves and twigs are food for a variety of wildlife and their buds were even eaten by native people. Cottonwood and willow leaves and buds contain salicylic acid, the main ingredient in aspirin.

The tall trees on the horizon of the cliffs are ponderosa pines. Pines from the eastern rim were logged, milled, and lowered by cable at the turn of the century (see page 50).

Cabins have front porches and fireplaces.

Zion lodge, designed by Gilbert S. Underwood, was built in 1925. Though it burned down in 1966, the lodge was quickly restored to its original rustic style. Today the lodge, with 80 rooms plus 40 historic cabins with fireplaces and front porches, provides excellent accommodations. The lodge includes a gift shop, snack bar, and restaurant. Internet access is available in the main lobby. Advance reservations for staying in the lodge are highly recommended.

FAST FACTS

> The name "Zion," which means "place of refuge," was given to the canyon by its first pioneer settler, Isaac Behunin.
> Trees have an ingenious design for "pumping" water up to their leaves.
> Dragonflies show amazing design and can even fly backwards.
> The Navajo Sandstone is one of the world's thickest single sedimentary rock units.

ZION LODGE

The Emerald Pools are a series of three pools below a small waterfall issuing from a high hanging valley. The trailhead to the pools starts at the footbridge across the road from Zion Lodge. The half-mile paved trail to the lower pool passes under an overhang where water cascades from the pool above (left). The middle pool is an additional half-mile on an unpaved trail. From there, it is a mile round trip to the upper pool. The Park Service map at the trailhead shows how to make this into a loop hike.

Question: Where did Igor Sikorsky get his idea for designing the helicopter?

C. Picnic lawn,
a good place to relax and take in
the high walls of Zion Canyon

D. Cottonwood tree,
which can lift over a thousand gallons
of water up into its leaves on a hot day

ELEVATION 4,230 feet

Mule deer were so named for their mule-like ears, which act like a satellite dish to collect the smallest noises of possible danger. "Muleys," as they are called, are very adaptable and can be found throughout all the western states. Probably the most visible animals in Zion, they are quite tolerant of human visitors. Occasionally they can be quite bold; so it's best not to approach them. Watch for them, especially at dawn and dusk, when they emerge from their sleeping spots to feed.

Female mule deer

A. The Narrows, a slot canyon over 2,000 feet deep

B. Virgin River meander, much smaller than the meander of the canyon, which was carved by much more water

NGELS LANDING

scurry about — but guard your lunch; the rascals will steal it. Remember, it is illegal to feed them.

Turkey vultures and golden eagles are often seen gracefully soaring in this area. The golden eagle's seven-foot wingspan is held flat while the turkey vulture's six-foot wingspan has an easily identified shallow "V" shape. Both have incredible vision enabling them to see miles beyond what the human eye can detect.

A bald-headed turkey vulture is designed without head feathers to clean up dead and decaying animals. Also, its digestive tract contains antibiotics to destroy the deadly bacteria; these same antibiotics continue to work in its excrement, sanitizing itself and its nearby environment.

Golden eagle

HIKING: Angels Landing is one of the most beautiful hikes in Zion National Park. Whether you trek to the summit or only hike partway, you will enjoy special places and make fascinating discoveries along the way. Access to the Angels Landing trailhead is from the footbridge across the road from The Grotto picnic area. This rewarding two and a half-mile hike (one-way) has a 1,500-foot elevation gain and is rated moderate to strenuous. The trail is five feet wide and paved for most of the first two miles, but narrows with steep drops near the top.

Nearly the entire Angels Landing Trail is on the Navajo Sandstone, the same rock as the walls of Zion Canyon. While hiking, stop to study the diagonal cross-bedding and long, flat bounding or truncation surfaces cutting across the cross-beds in the Navajo Sandstone. Are these dry, desert wind-blown sand dunes or underwater sand dunes (see page 142)?

Turkey vulture

After about one mile of uphill switch-backs, you enter refreshingly cool Refrigerator Canyon. It is a tall, narrow canyon where cooler-climate plants, such as big tooth maple, white fir, and Douglas fir thrive. Refrigerator Canyon is an elevated slot canyon that must have formed rapidly. The flowing stream likely did not carve the slot canyon. The most reasonable explanation for its formation is the rapid down-cutting by water during a catastrophic event, such as Noah's Flood (see page 144).

Walters Wiggles takes you up from Refrigerator Canyon to Scouts Lookout by 21 tight switchbacks, which incorporate a unique drainage system. The trail levels out and the pavement ends at Scouts Lookout,

THE GROTTO & AN

where hikers not comfortable with hiking to the top may stop and enjoy a great view.

Angels Landing summit is an additional half-mile up the ridge, steep on both sides with chain handholds, and is not for the faint of heart. From the summit, however, the view of Zion Canyon is spectacular. Notice the small, underfit meanders of the Virgin River compared to the large-scale valley meanders. This is a sign that much more water carved the canyon (see page 150).

ECOLOGY: Here on the rocky top of Angels Landing, enjoy Uinta chipmunks as they

Uinta chipmunk

The Grotto, located in a large, cool cottonwood grove, is a nice picnic area with tables, restrooms, and water. This is the trailhead hub for trails to Zion Lodge, the Emerald Pools, the West Rim, Scout Lookout, and Angels Landing. The Great White Throne evokes a sense of awe rising 2,800 vertical feet nearly directly above you to the east. The five-mile roundtrip hike to Angels Landing is spectacular, providing a 360º view overlooking Zion Canyon from 1,500 feet above the canyon floor.

FAST FACTS

> Angels Landing Trail is one of the top two attractions in Zion National Park, but you need to be in good physical condition.
> Slot canyons are narrow, steep-walled valleys with a catastrophic origin.
> Turkey vultures are part of nature's "clean-up crew."
> The stone building at The Grotto picnic area was the park visitor center until 1965.

The rock tower, on which Angels Landing sits, is reflected in the Virgin River (right). This photograph was taken by the United States Geographical and Geological Survey (now United States Geologic Survey) on the 1872 regional survey led by John Wesley Powell.

Question: What is an underfit stream and what does it have to do with the origins of Zion Canyon?

C. White Cliffs,
the third step of the Grand Staircase

D. Navajo Standstone,
forming a cliff up to
2,200 feet thick

THE GROTTO ELEVATION 4,297 feet
ANGELS LANDING ELEVATION 5,785 feet

The upper part of the trail to Angels Landing is often exposed, at times with over 1,000 foot cliffs on either side. Some of the steeper sections have a chain "rail" to assist hikers.

A. Navajo Sandstone, porous rock through which water seeps

B. Hanging garden, growing in the alcove of Weeping Rock that is always dripping water

calcium carbonate to form the travertine. It forms very quickly and is also seen as large mounds at the base of the springs.

ECOLOGY: Along Weeping Rock Trail, there is a great diversity of plants, ranging from bigtooth maple trees and maidenhair ferns to sage and cacti. This variety of plants is nurtured by the combination of moist and dry habitats. Interpretive signs along the trail identify many of these plants. The canyon wild grape growing here was useful and enjoyed by the pioneers as well as the Native Americans. Water-loving plants like the golden columbine and the vivid crimson monkey-flower are common.

Pacific maidenhair fern

Travertine deposits
around spring

HIKING: The trailhead for several hikes is just across the bridge. The left fork is the half-mile paved path to Weeping Rock alcove (pictured right). Signs along the trail identify many of the plants and trees of the area. The trails to Hidden Canyon and Observation Point also start here. Hidden Canyon is a two-mile round trip with a climb of 850 feet, part of which is along a sheer cliff with only chains for handrails, similar to Angels Landing. Hidden Canyon is a dry, narrow slot canyon that appears to have been cut and shaped by running water. The cool, shady, mile-long canyon is a favorite hike in summer. Observation Point is a strenuous eight-mile round-trip hike which climbs 2,148 feet and provides a spectacular view of Zion Canyon.

GEOLOGY: The spring which feeds Weeping Rock is the result of water percolating down through the porous Navajo Sandstone from the top of the plateau. When it reaches a less porous rock layer, often the underlying Kayenta Formation, the water is forced to flow laterally and ultimately out of the cliff face. Some studies estimate the water's journey from the top takes about a thousand years!

The alcove of Weeping Rock is formed by the seeping action of water, a process called sapping. However, Zion Canyon itself was not formed as the result of the present-day sapping we see at Weeping Rock, but the result of a catastrophic flow of

WEEPING ROCK

water, as evidenced by the large meanders seen in Zion Canyon (see page 150). Rapid sapping caused by a higher water content in the sediments at one time likely widened some side canyons. When viewed from a biblical worldview, the data suggests these features are better explained by the erosive power available during a global flood, like the Flood of Noah's day as described in the Bible.

The brown porous-looking rock around the springs is travertine. It is formed when water, which has been saturated with calcium carbonate as it flows through the rock, evaporates, redepositing the

Golden columbine

One of the most popular hikes in the Zion Canyon is the quarter mile Weeping Rock Trail leading up to a hanging garden of wildflowers, ferns, and mosses. Weeping Rock is a picturesque alcove where it "rains" every day of the year. A hanging garden is simply where plants grow on a vertical wall and are watered by springs seeping from the cliff. Many diverse plants are described along the trail.

FAST FACTS

> Along Weeping Rock Trail, cacti and ferns grow near each other due to a combination of wet and dry habitats.

> Weeping Rock water has taken hundreds of years to percolate down from the top.

> Golden columbine and red monkey flower are favorite hanging garden plants.

> The Native Americans built small "granaries" under overhangs to protect their grain from the critters.

WEEPING ROCK

From near the shuttle stop, one can see the remnant frame of a cable works just to the left of the highest peak on the eastern skyline. The cable way could lower bundles of lumber from the rim in about 2.5 minutes. The cable works delivered the lumber for the original Zion Lodge. In the early 1900s, this was more economical than wagon transport from northern Arizona.

Question: What formed these alcoves and side canyons?

C. Crimson monkey-flower,
growing on the wet cliff face

D. Travertine deposits,
formed by the buildup of calcium
carbonate as water evaporates

ELEVATION 4,310 feet

About 200 yards up the road from the
Weeping Rock shuttle stop is an alcove above
the road. In the white patch on the left side is
a small granary, probably built by the Anasazi
to store grains perhaps 1,000 years ago.

A. Angels Landing,
1,500 feet above the
canyon floor

B. Scout Lookout,
where the paved trail ends and
the narrow trail begins

HIKING AND CLIMBING: A half-mile walk north of the Big Bend bus stop is Menu Falls (upper left), so named because the menu at Zion Lodge once displayed the falls on the cover. The source of the falls is a spring at the base of the Navajo Sandstone.

Zion is the second most popular National Park for rock climbing, next to Yosemite. There are several "routes" up the 1,500-foot face of Angels Landing and this sheer cliff has been "free climbed" (meaning without ropes or protection). From March to June, climbing is banned in this area due to nesting peregrine falcons.

Menu Falls

GEOLOGY: How do vertical walls over 2,000 feet form? And why does the canyon start to widen so quickly below this point? One reason for canyon widening is because the softer Kayenta layer is exposed at the base of the Navajo Sandstone (seen on the east side of the road). The evolutionary literature says that as the river slowly erodes the Kayenta, it undercuts the sandstone, and "Rock slabs topple, cliffs landslide, and the canyon widens from here to the town of Springdale."[7] An alternative view is to envision this activity happening at the end of the Flood, with a catastrophic water flow rapidly eroding and removing the rock debris. The Park Service website lists several recent rockfalls in the canyon. If rockfalls had been going on at today's rates, after millions of years we should see huge aprons of rockfall debris at the base of these cliffs, but there is really very little (D).

We are told the river is constantly widening the walls of the canyon by erosion. Yet, from Big Bend south, the river is simply flowing over debris deposited by flooding, landslides, or an ancient lake; it is not cutting into the bedrock of the canyon. Catastrophic processes from only a few thousand years ago provide a better explanation for the formation of what is seen in Zion Canyon.

HISTORY: The Methodist minister, Frederick Vining Fisher, named The Great White Throne. As the story goes, during a trip with friends to the canyon in 1916, they caught a glimpse of the formation after

BIG BEND

storm clouds cleared. One of the boys cried "Oh, look! What is that? Name that Reverend!" After a moment of silence, Fisher responded, "Never have I seen such a sight before...boys, I have looked for this mountain all my life but I never expected to find it in this world. This mountain is The Great White Throne."[8]

ECOLOGY: Swallows and swifts often live in the same neighborhood as falcons, which prey upon them. While feeding their young, swallows may spend over 15 hours a day in the air; they feed, drink, bathe, and even mate while flying. They fly hundreds of miles a day to catch incredible numbers of insects. Both swifts and swallows fly south for the winter.

Big Bend offers majestic views of the towering peaks and the sheer walls of Zion Canyon. The Great White Throne is seen through the saddle just east of Angels Landing and is a spectacular view in the late afternoon light. Rock climbers can often be seen clinging to the cliffs and Angels Landing hikers can be seen on the ridge to the west. Big Bend gets its name from the large meander of the canyon.

FAST FACTS

> On an evolutionary timescale, there should be large amounts of rockfall at the base of these cliffs.

> Peregrine falcons can dive at speeds over 170 mph.

> Zion is the second most climbed National Park in the nation.

> The Great White Throne was named by Reverend Frederick Vining Fisher in 1916.

BIG BEND

The peregrine falcon nests on the cliffs of Zion Canyon and is considered to be the world's fastest animal, reaching speeds of 170 mph while diving for their prey. Their hunt usually consists of a high-speed mid-air collision with an explosion of feathers, in which the prey falls to the ground or is caught mid-air by the peregrine. At Zion you may see them near the cliffs looking for dinner.

Question: With frequent rockfalls recorded, if the canyon is millions of years old, where is all the rockfall debris?

C. Sheer cliff,
indicator of the rapid
carving of Zion Canyon

D. Lack of talus,
which should be piled high if this
canyon is millions of years old

ELEVATION 4,350 feet

*Great are the
works of the LORD…
Splendid and majestic
is His work, and His Righteousness
endures forever. He has made His
wonders to be remembered…*Psalm 111:2-4

The Great White Throne

A. **Pine tree,**
seemingly growing out
of sheer rock

B. **Iron staining,**
common in the Navajo
Sandstone

NAVA

eyelid for underwater sight, skin flaps over nostrils to keep the water out, and a high hemoglobin level for increased oxygen in their blood.

You'll likely see many inquisitive rock squirrels as you explore the canyon. They may beg for food but don't feed them; the most common injuries in the park are squirrel and chipmunk bites. Rock squirrels eat most anything: seeds, insects, berries, carrion, small birds, eggs, and roots. They are often seen, like the one pictured right, with their cheeks stuffed full of food. Their body is about a foot long with a speckled grayish-brown coat and a bushy tail. They live in colonies within burrows where they hibernate up to five months a year.

Rock Squirrel

GEOLOGY: The canyons of Zion National Park are carved in the 2,200-foot thick Navajo Sandstone. With its extensive cross-beds, this formation extends across most of the 130,000 square miles of the Colorado Plateau. This sandstone is consistent with the catastrophic deposits during the Flood and not that of a desert environment as often described in the evolutionary literature (see page 142).

The carving of these canyons starts once a channel is cut into its top surface by receding flood water, followed by continued rapid erosion forming a slot canyon. When the bottom of the slot canyon cuts through the Navajo Sandstone and into the softer less resistant rock of the underlying Kayenta Formation, canyon widening begins. Removal of Kayenta undercuts the Navajo Sandstone, causing collapse of unsupported slabs of the Navajo and the widening of Zion Canyon.

Evolutionary geologists believe the Virgin River eroded Zion Canyon in a few million years. But this would have produced a V-shaped canyon. The large scale of the canyon's meanders, indicate much more water cut the canyon (see page 150). Nor could this underfit stream have carved the many side canyons in the park. Note also the conspicuous lack of rockfall at the base of the cliffs. The catastrophic water flow necessary for carving a deep, vertically walled canyon, widening the canyon, and removing all of the eroded rockfall would be expected during the global Flood of Noah's time.

HIKING: The Narrows, a beautiful slot canyon, begins at the north end of the Riverside Walk, a mile upstream from the Temple of Sinawava bus stop. From there, the canyon becomes so narrow that

TEMPLE OF SINA\

hikers must take to the water between sections of trail. (Wading shoes and walking sticks can be rented in Springdale.) The full Narrows hike is 16 miles long and a backcountry permit is required. Check at the visitor center for permits and the latest weather forecast.

ECOLOGY: While walking the river, watch for a small bird called the American dipper, or water ouzel. They bob, or dip, when they walk or stand on rocks midstream. Amazingly, they fly over the river, then dive into the water and "fly" through the water or even walk along the bottom to catch aquatic insects. Special features of this amazing bird include an extra

American dipper

TEMPLE OF SINAWAVA

The Temple of Sinawava, set in a large beautiful rock amphitheater, is the last shuttle stop at the north end of Zion Canyon. The scene here is like an artist's painting with the majestic Pulpit and Altar rock spires presiding over the gently flowing tree-lined Virgin River. The Riverside Walk is a two-mile round trip hike on a paved trail to the gateway of The Narrows and access to the river's sandy shore. The Narrows is a tall slot canyon with surprises and wonders at every turn.

FAST FACTS

> The Temple of Sinawava was named after the coyote or wolf god of the Paiute Indians.
> Vertical walls, underfit streams, and lack of rockfall debris are key evidences for a recent and catastrophic canyon formation.
> The Narrows, with its hanging gardens, is a unique and delightful slot canyon.
> Watch for the little bird, the dipper, which dives into the Virgin River and "flies" underwater.

The Pulpit and Altar (right) standing in the middle of the Temple of Sinawava are erosional remnants of the Navajo Sandstone, likely left behind from the original carving of Zion Canyon.

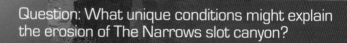

Question: What unique conditions might explain the erosion of The Narrows slot canyon?

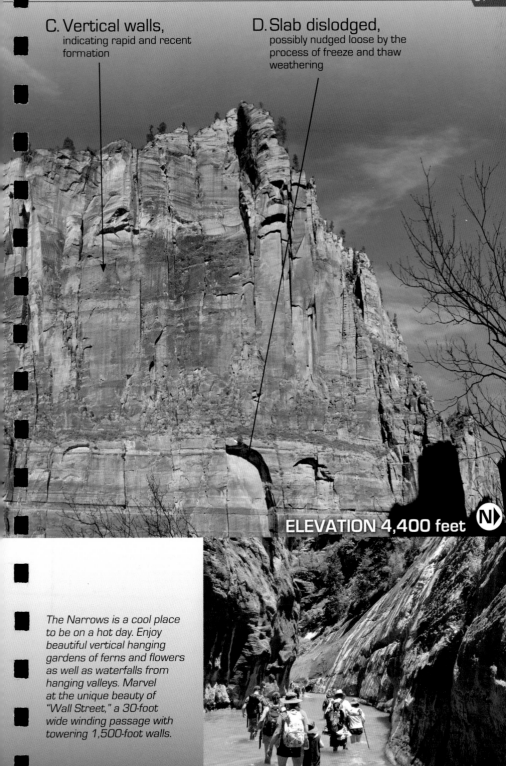

C. Vertical walls,
indicating rapid and recent formation

D. Slab dislodged,
possibly nudged loose by the process of freeze and thaw weathering

ELEVATION 4,400 feet

The Narrows is a cool place to be on a hot day. Enjoy beautiful vertical hanging gardens of ferns and flowers as well as waterfalls from hanging valleys. Marvel at the unique beauty of "Wall Street," a 30-foot wide winding passage with towering 1,500-foot walls.

A. Bridge Mountain, 6,803 feet above sea level, with Crawford Arch along the west side of the mountain

B. West Temple, 7,810 above sea level, the highest mountain in this part of Zion

OK

The upper and lower contacts of the Navajo Sandstone are flat at their boundaries with the Temple Cap Sandstone on top and the Kayenta Formation beneath. If erosion occurred between the deposition of these formations, there should be valleys and canyons at these contacts, but there are none, indicating that all three formations were laid down rapidly over a vast area (see page 146).

ECOLOGY: Condors and turkey vultures are often seen here soaring high above the canyons. Though both are scavengers in the vulture family, they do not compete for survival but rather cooperate. The turkey vultures can smell large dead mammals but are too weak to tear open the tough hide. The condor easily does this, eats what it needs, and leaves the rest for the turkey vultures (see page 161).

HISTORY: When started in 1927, the construction of the Zion-Mt. Carmel Highway and mile-long tunnel was considered an "almost impossible project." On the west side, the hairpin switchbacks climbed up the loose Kayenta Formation to the solid Navajo Sandstone where the tunnel was drilled and blasted. When dedicated on July 4, 1930, the tunnel was the longest in the United States.

HIKING: Canyon Overlook Trail, an easy half-mile hike, takes you to a large sandstone platform above The Great Arch (pictured from the highway below). The trailhead is across from a small parking lot at the east end of the Zion-Mt. Carmel Highway tunnel. Along the trail, a short metal walkway has low overhanging rock, so watch your head. The view from Canyon Overlook of Pine Canyon and the towering Navajo Sandstone walls is spectacular.

GEOLOGY: The Great Arch is actually an alcove. Alcoves are formed by water seeping into fractures (cracks). The water freezes, expanding the fractures and over time slowly pushing the rock apart. These portions then fall off in blocks, creating the alcoves.

From the tunnel eastward along the road, the beautiful swirls of red and white Navajo Sandstone, known as "slick rock," are evident. There are many places to park, walk out, and inspect the patterns. The diagonal lines, or cross-beds, are frequently separated by flat

CANYON OVERLO

bounding surfaces that planed off the tops of what once were large sand dunes. Research shows these surfaces were formed by underwater currents shearing off the tops of underwater sand dunes.[9]

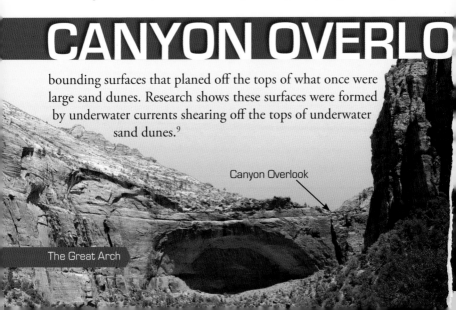

Canyon Overlook

The Great Arch

Canyon Overlook, perched on a mountain pass above the Great Arch, offers a breathtaking view west into lower Zion Canyon, surrounded by towering monoliths of red to tan Navajo Sandstone. In Pine Creek Canyon below, hairpin switchbacks climb up to the west entrance of the Zion-Mt. Carmel Highway tunnel. The mile-long tunnel carved in the Navajo Sandstone provides access to Checkerboard Mesa, the east park entrance, and the beautifully swirled layers of the Navajo Sandstone.

FAST FACTS

> The trailhead for Canyon Overlook Trail is at the east end of the tunnel.
> The massive Navajo Sandstone in Zion is over 2,000 feet thick.
> Pine Creek is a popular canyoneering adventure through a narrow slot cannon.
> Condors were reintroduced here after a successful inter-agency breeding program rescued them from near extinction.

CANYON OVERLOOK

California condors mature slowly, mate for life, and breed at six to eight years old. The female lays only one egg every two years. Adult wingspan is 10 feet. Researchers track them by radio, so you may see identification numbers and a small transmitter on their wings.

Question: Are the cross-beds in the Navajo Sandstone wind-blown desert sand dunes or underwater sand dunes?

C. Altar of Sacrifice, 7,505 feet above sea level, with vertical iron oxide stains

D. Pine Creek, too small to have eroded this deep, vertically walled canyon

ELEVATION 5,137 feet

Pine Creek, seen from the trail to Canyon Overlook (left), is one of the premier technical slot canyons in the Southwest. Rappelling into the canyon (right) requires experience, equipment, and a permit. This is a one-way six-hour adventure with multiple rappels, the last of which is over 100 feet directly into a cold pool. As you traverse Pine Creek Canyon, a spectacular display of light and shadow will reward your efforts.

C. Ponderosa pine,
common in this area

D. Towering cumulus cloud,
building upward often producing
afternoon thunderstorms

MESA

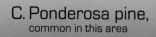

horn sheep

that shear off the diagonal cross-beds. Called bounding surfaces (A), some of these flat lines can be traced for miles. Many geologists admit these lines were formed by water action.[12] Research shows that slow water currents deposit sand into cross-bedded dunes; faster currents plane off the tops of the dunes.[9] There is no wind erosional process operating today that planes off the tops of a series of desert dunes without destroying the cross-beds. The regional extent of the sandstone, plus the distance of transport and the flat surfaces are all consistent with deposition during the Genesis Flood.

ECOLOGY: Visitors to Zion will probably see both desert cottontail rabbits as well as jackrabbits, which are technically hares. The larger jackrabbit is usually gaunt-looking with long ears and large hind feet. Coyotes and foxes have a hard time catching a jack if it notices them first, as jackrabbits can make an incredibly fast get away with 20-foot leaps at 40 mph.

The thin ears of hares are lined with many tiny blood vessels, which help radiate heat when it's hot. It would take a team of engineers to design a similar heat exchanger to be used in industry.

Jackrabbit

A. Bounding surfaces,
horizontal lines that divide
layers of cross-bedded
sandstone

B. Vertical cracks,
probably caused by expansion
during uplift of the area

CHECKERBOARD

GEOLOGY: The swirled and colorful sandstones between the tunnel and Checkerboard Mesa provide some of the most interesting scenery in Zion and the Southwest. Where not eroded away, the Navajo Sandstone originally covered 130,000 square miles of the Colorado Plateau. That is a huge amount of sand! Where did all that sand come from? The sand came from the north, as did most sandstones on the Colorado Plateau.[10] Some scientists think the sand had to come from as far away as Montana and Alberta, Canada, while others believe the sand came from the Appalachians about 1,600 miles away.[11]

Many evolutionary geologists believe the Navajo Sandstone is a hardened desert sand deposit, but a desert environment right in the middle of the Flood year would be a big problem for the Creation/Flood model. However, the majority of scientific evidence supports an underwater dune origin for the sandstone.

For example, consider that the top and bottom contacts of the Navajo Sandstone are flat, unlike any desert today. Also look at the flat, horizontal lines

Male big

Bounding surfaces

CHECKERBOARD MESA

Checkerboard Mesa is one of the most recognized features in Zion National Park. Its pattern is typical of the Navajo Sandstone, which is up to 2,200 feet thick. The Navajo Sandstone, with the Kayenta Formation below, forms the vertical cliffs in Zion, one of the tallest sandstone cliffs in the world. There are major scientific problems with the theory that the Navajo Sandstone was formed in a desert dune environment.

FAST FACTS

> The Navajo Sandstone provides convincing evidence of an underwater origin and some of the most beautiful scenery in the world.
> Zion Canyon was designated as a National Park in 1919, the same year the Grand Canyon became a National Park.
> Bounding surfaces in the sandstone suggest the Navajo Sandstone was deposited in a marine environment.
> Desert bighorn sheep were successfully reintroduced into Zion in the 1980s.

The "checkerboard" pattern of Checkerboard Mesa is caused by the dipping cross-bedded sandstone. The sandstone is divided into blocks by horizontal bounding surfaces and vertical weathering of cracks. The bounding surfaces represent the shearing off of the tops of the cross-beds by water, while the vertical cracks were probably caused by an expansion during uplift of the area.

Question: What is a bounding surface?

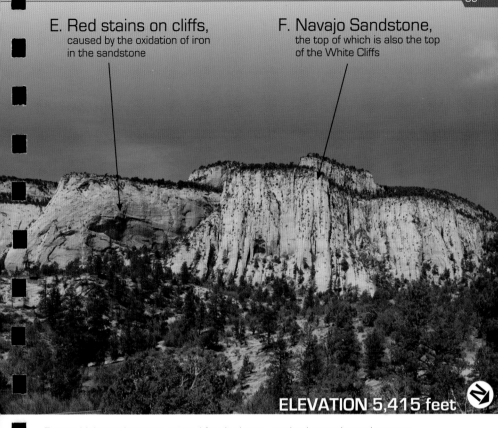

E. Red stains on cliffs, caused by the oxidation of iron in the sandstone

F. Navajo Sandstone, the top of which is also the top of the White Cliffs

ELEVATION 5,415 feet

Desert bighorn sheep are named for the large, curving horns the males grow (picture left). The females have horns also, but do not make a full curl like those of the male's. Bighorn sheep thrive in what appears to be the most inhospitable terrain — steep, rocky areas on the east side of Zion National Park. The courtship battle between bighorn rams in late fall is one of the most astounding sights in nature. Rising on their hind legs, opposing rams race towards each other, smashing their horns together with a terrific bang that sounds like a gunshot. Their lambs occasionally become prey for eagles, mountain lions, and bobcats, but few of their natural predators can match the adult's sure-footedness and agility.

Herd of bighorn sheep

A. The Sentinel,
7,157 feet above sea level

B. Kolob Terrace,
an erosion surface forming the
top of the White Cliffs

ECOLOGY: As you ascend to the rim country above Zion Canyon, you will gradually see quaking aspen appear until they make up large stands. These beautiful white-barked trees are responsible for the brilliant greens in spring and the magnificent golds in autumn. They are also one of the first trees to sprout after a fire because they can grow from the undamaged existing roots. Native Americans used the bark for a variety of purposes.

iffs   Kolob Terrace   Grey Cliffs

HIKING: The winding road to Lava Point (left) passes through a rural area with several beautiful horse ranches and wide meadows. There are several trailheads along the way providing access to Zion's backcountry. Many of these areas within the park require permits, available at the visitor center. For details, consult the park newspapers, *Zion Map & Guide* and the *Backcountry Planner.*

GEOLOGY: From Lava Point, one can see the upper three "steps" of the Grand Staircase. The bottom two steps (the Chocolate and Vermillion Cliffs) are not visible. The area in the foreground is the top of the third step, the White Cliffs (B), which forms Kolob Terrace. The mesas and canyons of Zion are carved into the Kolob Terrace. To the northeast, the Grey Cliffs are seen as the fourth step, with the Pink Cliffs, from which Bryce Canyon is carved, forming the top step.

From where you are standing, try to imagine all the sedimentary rocks from the Pink Cliffs on down stretching over you and to the south as far as you can see. The amount of sedimentary rock eroded off of the Grand Staircase was about 10,000 feet thick and represents

LAVA POINT

about 100 times as much material as is missing from Grand Canyon! From a creationist's perspective, the Grand Staircase was formed by the rapid "horizontal erosion" of a broad anticline, an uplifted dome of rock that stretched to the south over the Grand Canyon area (see page 140 and diagram on page 141). The evidence suggests this erosion had to be rapid (see page 123).

The top of the Kolob Terrace (B), formed by massive erosion of the sedimentary rocks above the White Cliffs, is seen in the foreground to the east. The Kolob Terrace is considered an erosion surface (see page 148) and is gently rolling with a few higher rock knobs sticking up (A). These rock knobs are erosional remnants of the upper strata and in geological jargon are called monadnocks.

Panoramic view of Kolob Terrace from Lava Point

Pink C

Lava Point provides a spectacular bird's eye view of Zion National Park, with a panoramic view of the Grand Staircase, including Zion's canyons and mesas. Lava Point is named after a layer of basalt (lava) that caps the sedimentary rocks at this location. To get to Lava Point, take the Kolob Terrace Road 20 miles north from the small town of Virgin, about 12 miles west of Springdale on Highway 9. Note the interesting erosional remnants of the Navajo Sandstone before reaching Lava Point.

FAST FACTS
> 10,000 feet of sedimentary rock has been eroded from this area forming the Grand Staircase.
> The top of the White Cliffs is an erosion surface.
> The erosion of Zion Canyon was from south- and west-directed currents coming from the eastern Grand Staircase area.
> Condors are often seen at Lava Point.

LAVA POINT

N

The trail to the famous, but out-of-the-way The Subway starts at the Left Fork Trailhead about eight miles from Virgin. The trail is a strenuous nine-mile round trip hike through forested terrain, across sandstone flats, and through a narrow canyon before reaching The Subway.

Question: What could have rapidly eroded the Grand Staircase?

C. Zion Canyon,
out of sight just beyond the mesas

D. The West Temple,
7,810 feet above sea level

ELEVATION 7,890 feet

Man has seen the beauty of this area since the Native Americans made it their home some 4,000 years ago. In the mid 1800s, Mormon pioneer Isaac Behunin gave the canyon the name Zion after the biblical place of refuge. Many of Zion's prominent features were given biblical names by Methodist minister Frederick Vining Fisher in 1916. Places such as the Court of the Patriarchs named after Abraham, Isaac, and Jacob, and The Great White Throne still hold those names.

May the Lord bless you from Zion,
He who made heaven and earth.
Psalm 134:3

Zion mesas,
pictured above

Bryce Canyon from Inspiration Point

From the visitor center going south, it is 18 miles to the final overlook at Rainbow Point. You will have ascended about 1,300 feet because the southern Paunsaugunt Plateau, on which you are driving, tilts slightly downward to the north. Since all overlooks are on the east side of the road, consider driving to Rainbow Point at the end and stopping at the overlooks on the return. In that way, you will always take a safer right turn to and from each overlook.

Actually, the first overlook is Fairyland Point and is accessed from outside the park, about one mile north of the park entrance. Many visitors miss this intriguing overlook, just a mile east of the main road.

SUGGESTED ITINERARIES

Begin your stay at the Bryce Canyon Visitor Center for information, museum exhibits, and the orientation film. Summarized below are a few recommendations.

Less than four hours: Ride the shuttle or drive your own vehicle to Sunset or Sunrise Point, and then take the shuttle to Bryce Point.

Half a day: Begin with the suggestions above, and then consider taking a short hike from one of the overlooks. There are several short hikes on segments of the Rim Trail, or take the Queen Garden Trail, the least strenuous trail which goes into the canyon.

DIXIE
NAL FOREST

Fairland Point
7758 ft
2365 m

FAIRYLAND CANYON

Fairland

BOAT MESA

Sinking Ship
7405 ft
2257 m

1 mi
2 km

CAMPBELL

Fairland Loop Trail

1 mi
2 km

Rim Trail

Tower
Bridge

CANYON

*Bristlecone
Point*

Visitor Center
7894 ft
2406 m

Sunrise Point

BRYCE   CANYON

Bryce Creek

Lodge

2 mi
3 km

Sunset Point

7200 ft
2195 m

North

No trailers
beyond
this point

Inspiration Point

Rainbow Gate
Road closed from
here to Rainbow Point
during winter storms

Rim   Trail

Bryce Point

Under-the-Rim Trail

2 mi
3 km

*Hat
Shop*

2 mi
3 km

3 mi
5 km

Paria View
8176 ft
2492 m

BRYCE CANYON NATIONAL PARK

6800 ft
2073 m

Yell
Group si

**Sheep Creek
Connecting Trail**

To Farview Point
(see large map)

BRIDGE

WILLIS

CANYON

Swamp Canyon
7998 ft
2438 m

6 mi
10 km

**Natural
Bridge**

PINK

**BRYCE CANYON
NATIONAL PARK**

**Swamp
Canyon
Connecting
Trail**

North 0   0.5   1 Kilometer

0   0.5   1 Mile

WHITEMAN

3 mi
5 km

*Swamp
Canyon Butte*
8302 ft
2534 m

**Agua
Canyon**

CLIFFS

AGUA

CANYON

CANYON

*Mud Canyon
Butte*
8330 ft
2539 m

SWAMP

Restrooms
Campground
Backcountry campsite
First aid
Picnic area
Emergency telephone

BENCH

3 mi
5 km

**Whiteman
Connecting
Trail**

*Noon Canyon
Butte*
8466 ft
2580 m

**Agua Canyon
Connecting Trail**

Ponderosa Canyon
8815 ft
2687 m

**Black
Birch
Canyon**

Iron
Spring

Under-the-Rim Trail

2 mi
3 km

**Piracy
Point**

Farview Point
8819 ft
2688 m

To Natural Bridge
(see inset)

Rainbow Point

**Yovimpa
Point**

Examine Everything Carefully

Bryce Canyon National Park

This section will help you examine the magnificent features of Bryce Canyon National Park in more detail as you stop at the overlooks and points of interest. There is a three-page layout for each of the major overlooks in the park. These will provide you with an understanding of what is seen, along with references to other sections for more detailed information. A variety of subjects are discussed, including the geology, ecology, hiking, and history. This park is at a higher altitude than most of Zion National Park, so the weather is cooler (see page 19). As a result, the plants and animals are generally different than in Zion National Park. The unique bristlecone pines are especially interesting.

There are services just outside the park entrance, including Ruby's Inn and Campground about a mile north of the park entrance. Just inside the entrance, there is a visitor center, where rangers are available to assist you. There are also two large campgrounds within the park as well as a lodge, restaurant, general store, showers, and laundry facilities.

The National Park Services newspaper, *The Hoodoo*, is available at the park entrance, the visitor center, and on the shuttle buses. The newspaper and this guide will be all you need to help make your visit to Bryce Canyon an extraordinary experience. For further ideas on how to plan your visit, read section three of this guide, *How to See Zion and Bryce Canyon*, on page 18. The overlooks are well marked but parking can be difficult in the peak of the season. The Park Service has a free shuttle bus that will take you to many of the overlooks.

Full day or more: Some of the most spectacular and lasting impressions of Bryce are seen from the trails. Take one of the combination loops starting at Sunset Point, a stroll along the rim, or a horseback ride into the canyon. Check at the visitor center for the schedule of free ranger programs. If you are staying overnight, get tickets at the visitor center for one of the park's astronomy programs. Consider getting up early to watch the sunrise over the canyon.

The following pages are designed to be used as you view each of the major overlooks in the park and to provide some things to examine and questions to ponder. As you view these overlooks, you will notice they provide slightly different views of the rim, hoodoos, and the canyon, as well as the Paria Valley and other plateaus to the east. However, viewing the different overlooks will help you understand how Bryce Canyon fits into the bigger picture of the Grand Staircase.

This park is etched into the Pink Cliffs, the top step of the five steps of the Grand Staircase. Also, you can see the Pink Cliffs, with its top layer of white limestone, in the higher Table Cliffs Plateau to the northeast across the Paria Valley. Uplift on a north-south fault in the Paria Valley has caused the Table Cliffs Plateau to move vertically upward about 2,000 feet. From the overlooks, you will catch glimpses of the 2,000-foot thick mass of volcanic rock on top of the northern Paunsaugunt and northern Table Cliffs Plateaus. The erosional pattern of the volcanic rock and the plateaus provides clues on how rapidly the area was eroded to form the stair-stepped pattern of the Grand Staircase (see page 123).

Enjoy this different perspective of Bryce Canyon National Park.

C. Queens Garden Trail,
starting at Sunrise Point and connecting to the Navajo Loop Trail

D. Queens Garden,
displaying a garden-like array of hoodoos

HIKING: Possibly the most picturesque hike in the park is the Queens Garden Trail. This mile and a half trail starts at Sunrise Point and descends a modest 320 feet into the canyon, winding its way through a garden of hoodoos. From a small spur trail, you can imagine Queen Victoria overseeing her garden below. You can also connect to the Navajo Loop Trail, which reaches the rim at Sunset Point for a total of about three miles.

Victoria hoodoo
any that bears
personalities,
r man made
tures.

ECOLOGY: The adaptable ponderosa pine growing here ranges from Canada to Central Mexico and from the California Coast to Nebraska. The older pines are recognized by their rusty-orange bark, split into big plates. When the bark is warmed by the sun, it gives off a pleasant aroma similar to vanilla. The seeds from the pinecones are a favorite food of the squirrels.

Notice the spindly limber pine standing on its roots off the south-west corner of the viewpoint area (right). Where was the ground level when this young tree sprouted from a seed?

Limber pine

A. Table Cliffs Plateau,
  capped by the White Member of the
  Claron Formation

B. Air quality
  can reduce visibility and obscure
  the view of distant plateaus

SUNRISE POINT

GEOLOGY: Bryce Canyon is carved in the Claron Formation on the Paunsaugunt Plateau which, in evolutionary thinking, was formed by an ancient lake that covered this area 50 million years ago. The Claron Formation is made up of two members. The lower Pink Member, 700 feet thick, is seen throughout the park and forms the Pink Cliffs of the Grand Staircase. The upper White Member is 300 feet thick and is only found at the higher elevations, mostly south of Sunrise Point. If you look northeast across the Paria Valley, these two members can be seen together along the edge of the Table Cliffs Plateau (A).

The Quee
is one of
the nam
animal
s

To the east, the waters of the Flood would have rapidly eroded the Paria Valley, leaving the edge of the Pink Cliffs exposed. The Pink Cliffs are composed mostly of soft limestone, which allows for rapid weathering, especially considering sapping, the seepage of water out of the saturated sediments. This would have been amplified by the high amount of precipitation of the Ice Age.[13] Since all of North America could be reduced to sea level in less than 40 million years at the present erosion rate (see page 146), the evidence suggests that Bryce Canyon is a relatively young feature.

SUNRISE POINT

Sunrise Point, the first stop after the visitor center, offers a commanding view of the Paria Valley to the east and the Aquarius and Table Cliffs Plateaus in the distance toward the northeast. This valley represents rapid erosion during the carving of the Grand Staircase. The limestone of the Claron Formation is softer than some other limestones in the region, which allowed Bryce Canyon's hoodoos to be eroded rather quickly, likely after the receding of the Flood.

FAST FACTS

> The Claron Formation was named the "Pink Cliffs" by Clarence Dutton in 1870.

> Ponderosa pine is the most common evergreen tree at this location.

> Standing on its bare roots, the limber pine shows the amount of erosion that has occurred in its short lifetime.

> Post-Flood erosion has been rapid, leading to the conclusion that Bryce Canyon is young.

Horse or mule rides can be taken down into the canyon. Wranglers lead two-hour and half-day trips. Inquire at Bryce Canyon Lodge or contact Canyon Trail Rides in the town of Tropic, just east of the park.

Question: Why would rapid erosion rates in Bryce Canyon suggest its landscape is young?

E. Bryce Point,
capped by conglomerate, which forms thin layers in the Pink Cliffs

F. Sunset Point,
only half a mile walk south along the rim

ELEVATION 8,015 feet

This is the view from the bottom of Queens Garden Trail. The arrow shows the trail above, which is halfway down the trail from the rim.

C. Canaan Mountain,
capped by Claron Formation limestone

D. Bryce Creek,
a drainage likely carved out after the Flood, especially by summer flash floods

Along the trail, evergreen Douglas firs, 500- to 700-years old, stand in sharp contrast to the Pink Cliffs. This trail can be combined with the Queen's Garden Trail to form a longer loop ending at Sunrise Point.

ECOLOGY: The white-throated swift and the violet-green swallow both live in this neighborhood and are preyed upon by peregrine falcons. If not taken by surprise, they have both been seen to out-maneuver the falcons. To do this, swallows have 1,200 tiny muscles attached to the roots of their wing feathers, which can shift the load-bearing surface of their wings in a split second. Isn't it hard to believe those features evolved from reptile scales, rather than the designed handiwork of the Creator?

Even though the white-throated swift is about an inch bigger, the two birds are difficult to tell apart. The swifts are basically black and white with jerky flight similar to bats. Swallows fly smoothly, have brownish wings, and, though often looking black, in the right light they are a brilliant iridescent violet-green, hence their name.

rail descends treet from Point (F)

5" violet-green swallow

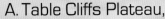

A. Table Cliffs Plateau,
capped by the White Member of the
Claron Formation

B. Thor's Hammer,
seen below to the north

SUNSET POINT

GEOLOGY: Hoodoos, which come in various shapes and sizes, are caused by the different weathering properties of rock in the Pink Cliffs. Most hoodoos have more resistant rock in the upper layers, protecting the softer limestones underneath and delaying their erosion. This results in the steep vertical formation. The hoodoo is then easily fragmented by erosion. Intermittent precipitation and the fact that the temperature passes through the freezing point 200 times a year in Bryce, results in great freeze-thaw weathering, loosening blocks of rock. Sometimes, when the morning sun warms a southern exposure after a freezing night, rocks come crashing down.[14]

Sunset Point offers views of the some of the most famous hoodoos in the park. Directly below to the south is the Silent City, forming a maze of spires and fins. On the northern edge of the overlook stands the solitary Thor's Hammer (B and far right).

HIKING: The Navajo Loop Trail is the most popular hike in the park. It descends from Sunset Point to the bottom of the canyon where it passes through the slot canyon called Wall Street (F and above).

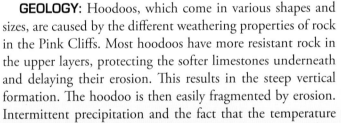

Navajo Loop
into Wall S
Sunset

N

Sunset Point is about half a mile south of Sunrise Point. They are connected by the Rim Trail. Thor's Hammer, the most photographed hoodoo in the park, is located just below Sunset Point to the north on the Navajo Loop Trail, which begins and ends here. The south leg of the trail descends through a series of switchback to the famous Wall Street (F).

FAST FACTS

> Hoodoos are formed in the Pink Cliffs by the effects of freeze, thaw, wind, and rain erosion on rock layers of variable hardness.
> White-throated swifts and violet-green swallows live among these cliffs.
> The Navajo Loop Trail guides hikers through some of the most interesting geological wonders below the rim.
> Large, tall Douglas firs grow in the sheltered environment of the Wall Street slot canyon on the Navajo Loop Trail.

Bryce Canyon Lodge, with its surrounding cabins (below), is next to Sunset Point. They were constructed in several phases in the 1920s.

Question: What causes hoodoos?

E. Inspiration Point,
one of the most commanding
views of Bryce Canyon

F. Wall Street,
entrance to a narrow slot canyon
on the Navajo Loop Trail

ELEVATION 8,000 feet

Thor's Hammer, shown clockwise at
sunrise, mid-day, late afternoon, and
sunset, displays how the colors of the
canyon change minute by minute. The
camera can catch but a glimpse of the
splendor found in these canyons.

C. Aquarius Plateau, volcanic rocks on the northern Table Cliffs Plateau

D. Table Cliffs Plateau, topped by a white layer that overlies a pink layer, all part of the Claron Formation

The northern Paunsaugunt Plateau (A), visible in the distance north of Inspiration Point, has dark volcanic rocks overlying the light-colored Claron Formation. The volcanic rocks once extended south over this area, covering the southern Paunsaugunt Plateau. Mysteriously, the hard volcanic rocks have been eroded northward, but the underlying soft Claron Formation was hardly eroded at all. A slow process of erosion over millions of years does not explain this. The only way to explain the lack of erosion of the soft Claron is if the rapid erosion stopped after eroding away the volcanic rocks. This would be expected at the end of the runoff stage of the Flood. Similar events happened on the Aquarius Plateau (C) and Table Cliffs Plateau (D) to the northeast (see page 123).

, just one pher's ors

ECOLOGY: Pocket gophers are named for their external cheek pouches used to transport food, but have no direct opening to their mouths. They spend most of their life underground, continually tunneling through soils to eat, gather, and store roots and leaves. Because they are active year round, the pocket gopher is always an "in season" meal for predators like coyotes, bobcats, foxes, owls, and snakes.

Northern pocket gopher

A. **Volcanic rocks**
on top of the Claron limestone
of the northern Paunsaugunt
Plateau

B. **Navajo Loop Trail,**
descending through Wall Street
from Sunset Point

INSPIRATION POI

Great horned ow
of the pocket g
many preda

GEOLOGY: Bryce Canyon was carved from the southeastern rim of the Paunsaugunt Plateau, creating the bowl-shaped canyons like Bryce Amphitheater (left portion of panorama) seen below Inspiration Point. The striking pink, tan, and red Claron Formation is the perfect medium for this artistic work of erosion by rain, freeze, thaw, wind, and running water. These erosive forces continue to sculpt hoodoos from the fins (see page 102).

Evolutionary scientists believe the Claron Formation formed in an ancient lake called Lake Claron. They propose that limestone was predominantly deposited in this lake over millions of years. But it does not seem likely that so much limestone can be deposited in a lake; limestone is mostly formed in saltwater. Besides that, in-flowing rivers would bring in sand, mud, and rocks as well. Such a proposed lake should contain certain fossils and alkali deposits, but there are none. A worldwide flood better explains the deposition of over 2,000 square miles of limestone, over 1,000 feet thick, with hardly any fossils.

INSPIRATION POINT

From Inspiration Point, beautiful and unusual hoodoos, sculptured walls, and deep gullies can be seen below in Bryce Amphitheater. Catch the fascination of Bryce Canyon with a hike in the canyon or a walk along the rim. To the northeast, you see volcanic rocks capping the Aquarius Plateau (C) just like the volcanic rocks capping the northern Paunsaugunt Plateau (A), to the north. Mysteriously, these hard rocks eroded while the underlying soft Claron Formation failed to erode.

FAST FACTS

> Narrow sculptured ridges, called fins, stand between the deeper gullies carved by running water from precipitation.

> Mountain lions catch their prey 80 percent of the time.

> Carrying food in their cheek pouches allows pocket gophers to escape their predators with no need to drop their food.

> The Bryce Canyon paintbrush (below) was first discovered at Inspiration Point.

The Bryce Canyon paintbrush, first found here, is a semi-parasitic plant making part of its food by photosynthesis while also stealing food from the roots of nearby host plants. The red is not actually the flower but a modified leaf called a bract; the actual flower is inconspicuous and without scent. Their nectar is an important food source for hummingbirds, which have almost no sense of smell but are attracted to the bright red color.

Question: Was the Claron Formation deposited during a global flood or in an ancient freshwater lake?

E. Canaan Peak,
an erosional remnant capped by
the Claron Formation

F. Fins
eroded plateaus that over time
form into hoodoos [see page 102]

ELEVATION 8,100 feet

Mountain lions inhabit the park, but are rarely seen, preferring to avoid people.
Mountain lions have as many as 30 different names, including cougar, panther,
puma, painter, catamount, leon, etc., all referring to the same animal. Skilled hunters,
mountain lions have one of the highest success rates of any predator, catching their
prey as much as 80 percent of the time. They prefer larger and harder-to-catch prey
such as deer and can consume about one mule deer per week. They also hunt rodents,
coyotes, and birds, and will often hide the leftovers from a kill for later eating.

Mountain lion and cub

C. Volcanic rocks
on top of the northern
Paunsaugunt Plataeu

D. The Alligator,
whose harder layer on top is
more resistant to erosion and
holds the fins together

west wall of
Point

The small grottos, or caves, seen along the rim (A and left), are formed by sapping, the process in which rainwater and snowmelt trickle down into the sandstone. As the water spreads horizontally and out of the edge of the cliff, it dissolves portions of the rock, forming the cave-like grottos. Notice how the grottos appear in a row, as they are formed in the same horizontal layer.

ECOLOGY: A member of the mustard family, the western wallflower blooms in early summer throughout the drier regions of the western United States. In fact, it has the widest distribution of any flower in Utah, growing at elevations between 2,500 to 12,500 feet. Watch for it in June near the trails, viewpoints, and along the main road.

HIKING: The three-mile Peekaboo Loop Trail in the canyon bottom offers spectacular scenery, including the Peekaboo Box Canyon and the Wall of Windows. This is also a wonderful horse trail, accessible from the corral at Sunrise Point. The Bryce Canyon Hiking Map, available at the visitor center, offers information for this and other hikes in the park.

Peekaboo Loop Trail

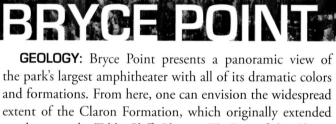

A. Grottos, caused by sapping, the seepage of groundwater through a sandstone layer

B. Bryce Amphitheater, made up of several smaller amphitheaters stretching from Bryce Point to Sunset Point

BRYCE POINT

Grottos on the Bryce

GEOLOGY: Bryce Point presents a panoramic view of the park's largest amphitheater with all of its dramatic colors and formations. From here, one can envision the widespread extent of the Claron Formation, which originally extended northeast to the Table Cliffs Plateau (F). Part of the Claron Formation, the Pink Cliffs, extends dozens of miles to the north and west, and is the formation into which Cedar Breaks National Monument and Red Canyon are carved. Other formations of the Grand Staircase, such as the Navajo Sandstone and the Shinarump Conglomerate, are even much larger than the Claron Formation. Such widespread deposition of sediments is not happening on this scale today but would be expected during the Genesis Flood (see page 145).

The dark volcanic rock on top of the northern Table Cliffs Plateau is 2,000 feet thick and called the Aquarius Plateau (E). This same volcanic rock also occurs just north of Bryce Canyon on top of the Paunsaugunt Plateau (C) (see page 142).

Western wallflower

BRYCE POINT

Famous for its magnificent sunrises, Bryce Point provides one of the most inspiring vistas of the Bryce Amphitheater to the north. As the sun rises, it is like a fire engulfing the canyon to overwhelm the shadows, producing an amazing display of colors. Bryce Canyon gets its name from Ebenezer Bryce, who settled in the Paria Valley east of the canyon in 1870. Notice the numerous small erosion-caused, cave-like grottos which occur to the west just below the rim.

FAST FACTS

> The Claron Formation, forming the Pink Cliffs, covers over 2,000 square miles and is over 1,000 feet thick.
> Other formations of the Grand Staircase cover tens of thousands of square miles.
> The western wallflower has the widest distribution of any wildflower in Utah.
> The grottos are formed by water seeping out of the rock, a process called sapping.

Notice the light brown color of the rock at the top of Bryce Point (right). Part of the Claron Formation, this brown conglomerate rock contains sand and small pebbles. It outcrops sporadically along the rim of Bryce Canyon. The pebbles are a type of rock not found anywhere near Bryce Canyon (see page 143).

Question: What could have caused such widespread layers of rock?

E. Aquarius Plateau, volcanic rock on the northern Table Cliffs Plateau

F. Table Cliffs Plateau, with the White and Pink Members of the soft Claron Formation

ELEVATION 8,300 feet

The colors of the Claron Formation are caused by oxidation, the reaction of certain minerals with oxygen. The red, pink, yellow, and brown colors are shades of iron oxide. The purple hues are oxidation of the element manganese.

C. Large hoodoo,
capped by the White Member of
the Claron Formation

D. Yellow Creek Spring,
a small spring flowing into
Yellow Creek

...t on the hoodoo
...e Canyon (C)

crickets chirping, count the number of chirps in 13 seconds, add 40, and the result will be very close to the degrees Fahrenheit.[15]

Peregrine falcons are sometimes seen soaring along the rim of Paria View and may even be seen nesting in the cliffs. Peregrines are identifiable by their gray back, black and white head, and pointed, sickle-shaped wings (see page 54).

Mountain short-horned lizards may be seen in dry and rocky areas of Bryce Canyon. Often called "horned toads," short-horned lizards are harmless, but they do have some interesting means of defense. Like chameleons, horned lizards can take on the color of the soil around them — creating an effective camouflage. If caught by a predator, the adults puff up and poke their attackers with sharp little spikes. And some of them can squirt blood up to five feet out of the corner of their eye. If you picked one up and it squirted blood at you, wouldn't you be startled enough to let it go?

Mountain short-horned lizard

A. White Member
within the Claron Formation

B. Pink Member
of the Claron Formation,
forming the Pink Cliffs

PARIA VIEW

GEOLOGY: The White Member of the Claron Formation (A) is distinctly seen here, overlying the Pink Member (B). Patches of conglomerate, which contain rocks rounded by being transported in water, occur within the Claron Formation. The type of rock in the conglomerate does not outcrop in the vicinity of Bryce but comes from around eastern Nevada (see page 143).

Looking southeast down the valley, the edge of the White Cliffs (E) emerges. Several north-south canyons are carved into the edge of the White Cliffs. This erosion likely occurred when the Paria Valley was rapidly eroded during the Flood by a channelized current flowing south. Zion Canyon is also carved vertically out of the edge of the White Cliffs. The edge of the tall cliffs of the Grand Staircase are noted for vertical-walled canyons cut perpendicular to the line of the cliffs, such as seen north of Kanab, Utah.

ECOLOGY: How hot is it? Did you know you can tell about how hot it is by listening to crickets? When you hear

Evening lig
of Bry

Snow tree cricket

Paria View is capped by the White Member of the Claron Formation, which outcrops sporadically in Bryce, but can be seen across the Paria Valley in the Table Cliffs Plateau (see bottom of pages 111 and 112). Channelized erosion in the Paria Valley to the east carved several north-south canyons along the edge of the White Cliffs seen from here.

N

FAST FACTS

> Paria is a Paiute word meaning "water with elk" or "water with mud."
> The origin of the Claron Formation is a mystery for evolutionary geologists.
> The chirping of crickets can tell us how warm it is.
> In winter, a cross-country ski trail is the only access to Paria View.

PARIA VIEW

Pronghorn antelope are often called "speed goats" by local ranchers. They are the second-fastest land mammal in the world. Pronghorns reach speeds up to 60 mph, usually leaving their predators in the dust. Only the cheetah can run faster.

Question: Why do short-horned lizards squirt blood out of their eye?

E. White Cliffs,
with deep canyons eroded into them, such as at Zion

F. Erosion surface
on top of the Navajo Sandstone that forms the White Cliffs

ELEVATION 8,175 feet

Pronghorn antelope may be seen along the road to Paria, especially in late spring and early summer. They like to leave the sagebrush prairies to the north of the park to give birth to their young in the seclusion of the forest.

A. Ponderosa pine,
common in the west and
at Bryce Canyon

B. A fin,
a ridge from which further
weathering produces
windows and hoodoos

The spadefoot toad is another remarkable animal that lives in Swamp Canyon. This little toad has a "spade" on its hind feet that enables it to dig as much as 12 feet down into the sand and lie dormant, sometimes for almost a year. They instinctively know to emerge when a major thunderstorm causes pools to form. It croaks for females, which may come from as far as a mile away to mate and lay eggs in the pools. In as little as nine days, the eggs hatch to tadpoles, which transform into toads. The young toads have, at most, a few weeks to eat enough food to survive before burying themselves down in the sand to wait for the next rainstorm.

Spadefoot toad

ECOLOGY: Bryce Canyon National Park was established to protect and better understand the unusual and beautiful rock spires called "hoodoos." But later, as overgrazing and predator extermination took their toll on the region, Bryce also became a small but critical refuge for scores of animal species, including everything from the endangered Utah prairie dog to the elusive mountain lion.

In the valley below, two tiny springs make Swamp Canyon a "wetland" that sustains lush vegetation like willows, grasses, and flowering herbs. Tiger salamanders and other animals thrive here since the canyon remains wet year round.

Swamp Canyon Trail is a favorite for bird watchers, since it transverses four distinctly different habitats. Thus, it offers the opportunity to see much of Bryce Canyon's great diversity of flora and fauna without a long backcountry hike. The most commonly seen animals in Swamp Canyon are listed by their four local habitats:

- Rim — with Clarks nutcrackers, Steller's jays, and western tanagers
- Canyon — with violet-green swallows and short-horned lizards
- Meadow — with prairie dogs and mourning doves
- Prairie — with jackrabbits and cottontails

Mule deer, squirrels, chipmunks, turkey vultures, ravens, nighthawks, and dark-eyed juncos are seen in all four habitats.

The Missouri, or Rocky Mountain iris, is both beautiful yet deadly. The leaves have a high concentration of the chemical irisin, which is poisonous to livestock and people alike. Symptoms can become extreme and include severe and simultaneous vomiting and diarrhea. This flower prefers

Missouri iris

SWAMP CANYON

moist soils and has very beautiful blue to violet flowers with a bit of deep yellow running down the center of each flower petal. The leaves are sword-like.

The tiger salamander, sometimes a foot long, inhabits meadows, forests, shady canyons, and springs. Active at night to avoid predators, their diet includes earthworms, large insects, small mice, and amphibians. The female lays eggs that go through a larvae stage before transforming into a four-inch sala- mander about a month later. Adults live up to 25 years.

Tiger salamander

Swamp Canyon is small and sheltered compared to the other overlooks. The view is of a narrow canyon bordered by fins and hoodoos. There is no actual swamp here, but compared to the rest of the park, Swamp Canyon is a virtual wetland, with some plants that grow nowhere else in the park. Swamp Canyon connects with the Under-the-Rim Trail and a four and a half-mile loop trail, which is enjoyed by nature lovers. The trail is a great place for bird watching.

FAST FACTS
> Two tiny springs make Swamp Canyon a "wetland" year round.
> Swamp Canyon flora and fauna live in four local habitats.
> Over 170 different birds have been seen in Swamp Canyon.
> In Swamp Canyon, the peculiar spadefoot toad may be heard croaking after a summer thunderstorm.

SWAMP CANYON

The western tanger is a colorful little bird that spends its summer in Bryce and other western forests. The male has a red head with bright yellow body and black wings, back, and tail. The female is without red, and has a duller yellow and gray color. Tanagers eat some fruit and berries but mostly feed on insects, often catching them in mid-air like swallows do. Despite his bright colors, the male is fairly inconspicuous, spending most of his time in the trees.

Male western tanager

Question: Why are there four local habitats with a variety of plants and animals at Swamp Canyon?

C. Swamp Canyon, with four distinct habitats and a great place for bird watching

D. Small mesa, an erosional remnant in the Pink Cliffs

ELEVATION 7,998 feet

The Utah prairie dog is tawny to reddish-brown in color, with a short white-tipped tail and black "eyebrows." They are very social and live in burrows in large colonies. Utah prairie dogs hibernate in the winter and in early March, males emerge from their burrows with the females coming out later in the month. Mating is complete in early April and the "pups" are born in late April or early May. The playful juvenile prairie dogs emerge from the den after about six weeks and attain adult size by October.

Utah prairie dog

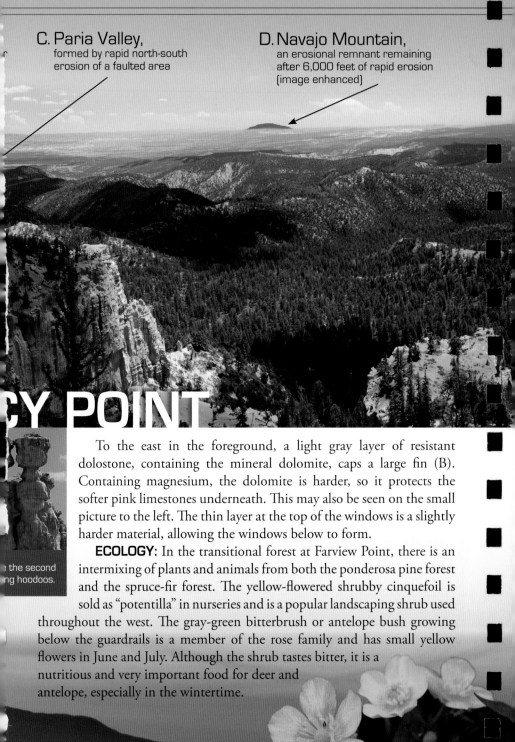

C. Paria Valley, formed by rapid north-south erosion of a faulted area

D. Navajo Mountain, an erosional remnant remaining after 6,000 feet of rapid erosion (image enhanced)

CY POINT

To the east in the foreground, a light gray layer of resistant dolostone, containing the mineral dolomite, caps a large fin (B). Containing magnesium, the dolomite is harder, so it protects the softer pink limestones underneath. This may also be seen on the small picture to the left. The thin layer at the top of the windows is a slightly harder material, allowing the windows below to form.

ECOLOGY: In the transitional forest at Farview Point, there is an intermixing of plants and animals from both the ponderosa pine forest and the spruce-fir forest. The yellow-flowered shrubby cinquefoil is sold as "potentilla" in nurseries and is a popular landscaping shrub used throughout the west. The gray-green bitterbrush or antelope bush growing below the guardrails is a member of the rose family and has small yellow flowers in June and July. Although the shrub tastes bitter, it is a nutritious and very important food for deer and antelope, especially in the wintertime.

the second ng hoodoos.

Cinquefoil

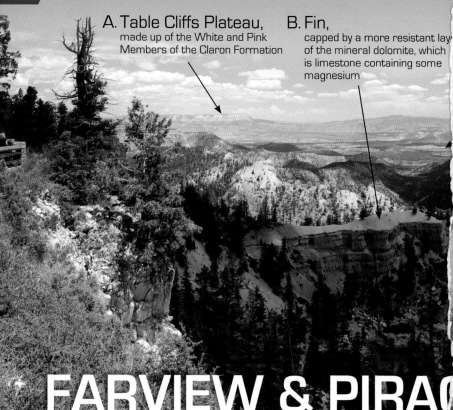

A. Table Cliffs Plateau,
made up of the White and Pink
Members of the Claron Formation

B. Fin,
capped by a more resistant lay
of the mineral dolomite, which
is limestone containing some
magnesium

FARVIEW & PIRA

GEOLOGY: On a clear day, Navajo Mountain (D), standing 10,388 feet above sea level, can be seen near the Arizona border about 90 miles to the southeast. While this volcanic mass was being formed, it never made it to the surface, but rather was intruded into the surrounding sedimentary rock layers. Yet it now stands about 6,000 feet above the surrounding terrain of the Kaiparowits Plateau. This means that a huge amount of sedimentary rock has been eroded from around it. Based on the height of the Grand Staircase and the slight northward tilt, the amount of erosion was about 10,000 feet.

This erosion must have been rapid, or else Navajo Mountain would not be left standing. If erosion of the sedimentary rocks was slow over millions of years, the volcanic rock would have also eroded. Since erosion of mountains is faster than erosion on a rolling plateau, Navajo Mountain should have been quickly destroyed. Thus, it is an erosional remnant left after rapid erosion of the surrounding sedimentary rocks.

Windows ar
step in form

Sunset behind Navajo Mountain

Farview Point is a good location to visualize the stages of hoodoo formation due to weathering (see below). The forest here is in a transition zone between the open ponderosa pine forest and the spruce fir forest found in the higher elevations. Piracy Point is a separate viewpoint a short quarter-mile walk to the north of Farview Point and provides a different perspective of the canyon below.

FAST FACTS

> Navajo Mountain indicates rapid erosion of the sedimentary rocks south of the Grand Staircase.
> The mineral dolomite is more resistant to erosion than limestone.
> The two different types of striped squirrels at Bryce are easy to tell apart.
> Differential erosion is the foundational process in creating hoodoos.

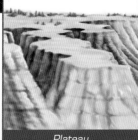

Plateau

Hoodoo formation usually begins with a narrow fin of rock. Rocks are removed by freezing and thawing to produce windows in the fin. As the windows grow they can become large enough to be called arches. Eventually the window or arch collapses and the two broken legs become hoodoos.

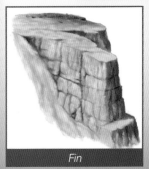

Fin

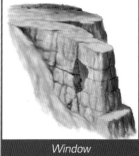

Window

Hoodoo

Question: What element in dolomite makes it a more resistant capstone?

E. Eroded valley,
whose exposed southern
extension is the Grey Cliffs

F. Rainbow Point,
the southeast tip of the park and
of the Paunsaugunt Plateau

ELEVATION 8,819 feet

Practically every visitor to Bryce will see squirrels. The most commonly seen species at Bryce is the golden-mantled ground squirrel, although there are plenty of rock and red squirrels as well. Golden-mantled ground squirrels are often confused with chipmunks, but they are easy to tell apart — the chipmunks have stripes on the sides of their faces and the squirrels do not.

Chipmunk

Golden-mantled ground squirrel

DO NOT FEED
WILDLIFE!
$100 PENALTY
Please report
violators

A. Natural Bridge,
technically an arch, since no
stream ever ran under it

B. Bridge Canyon,
seen through the arch

The canyon wren and the rock wren are both inhabitants of Bryce. Their flat body shapes make it possible to creep under rocks to find bugs to eat. Wrens are often mistaken for a mouse at first because people don't expect to see a bird creeping under and around rocks. Both wrens are only about five inches long, the size of small sparrows, but the rock wren is gray with a dark breast, not reddish-brown like a canyon wren which has a white breast. The rock wren is very trustful and if you move slowly, you can get within a few feet of it.

Canyon wren

GEOLOGY: Natural bridges and arches are fascinating and appear to be solid and enduring. This is because we see them in the present. But over time, the structures are really unstable. The effect of weathering increases the opening and reduces the overhanging rock. A rock arch or natural bridge may last hundreds to thousands of years, but eventually it will collapse.

Practically all the arches in Bryce Canyon National Park would have formed in a relatively short time by normal erosion processes because the rocks are generally soft and weather easily. Small arches and natural bridges like these would be expected to form after the Flood by normal weathering and erosion.

However, the freestanding arches, those that are narrow and supported at each end, are more mysterious. Many freestanding rock arches are on ridges or the sides of a ridge or mountain, like Delicate Arch in Arches National Park and Crawford Arch in Zion National Park (see page 35). In evolutionary thinking, these types of arches are expected to form slowly over many tens of thousands of years by weathering and erosion. We observe the collapse of freestanding arches today, such as Wall Arch in Arches National Park in 2008, but we do not witness the formation of freestanding arches. It seems that freestanding arches require rapid erosion on ridges or mountains or along the edges of these obstacles, which could easily have occurred during the receding of the Flood but not by present processes of erosion over a long time.

ECOLOGY: Did you know coyotes hunt cooperatively with badgers? The Navajo have legends about the coyote and badger hunting

NATURAL BRIDGE

together, and scientists have found that coyotes hunting with badgers can catch about one-third more rodents than when hunting alone.[16] A coyote is good at chasing a rodent down a hole, but not very good at digging it out. So he gets a badger to dig it out while he stands guard. If the rodent tries to escape from another hole, the coyote chases it down again. The badger may dig eight feet down to catch the rodent, which he shares with the coyote. How do cooperative systems like this evolve, if competition, not cooperation, is the driving mechanism of "survival of the fittest" evolution?

Badger

Natural Bridge overlook is named for what is called a natural bridge. But as the interpretive sign states, it is misnamed; it is really an arch (see page 147). A natural bridge is formed by a stream that undercuts the rock, usually in a meander. Arches are formed by more subtle erosional effects and can be of several types. There is no active stream or any evidence that a stream ever existed under an arch.

FAST FACTS

> Arches and natural bridges are unstable and eventually collapse.
> The arches in Bryce Canyon most likely formed after the Flood.
> Freestanding arches, like those seen in Arches National Park, are different and require rapid formation.
> It is amazing that coyotes and badgers will occasionally cooperate to hunt rodents together.

NATURAL BRIDGE

The deer mouse is a cute little animal with big eyes and big ears. Its head and body are about 3 inches long, and the tail adds another 3. They range in colors from gray to reddish brown with a white underbelly. Unfortunately they are one of many rodents in the southwest who carry the Hantavirus, a rare respiratory disease. People contract the virus through direct contact with rodents and rodent droppings. Producing flu-like symptoms, Hantavirus can be deadly if untreated.

Question: Why are there freestanding arches today?

C. Arch formation, with its softer layer below eroded away, leaving the harder cap rock above

D. The Pink Member of the Claron Formation

ELEVATION 8,627 feet

Because of slightly different minerals, the hardness of the Pink Member of the Claron Formation varies from layer to layer and within a layer. So, during the weathering processes that form hoodoos, different rates of weathering produce all kinds of shapes. In the picture to the left, the crown on top of the pedestal weathered slower because it is harder, while its contact with the pedestal was undercut faster because the softer rock there is easier to erode.

C. The Hunter,
eroded to the point it no longer looks like a hunter

D. The Rabbit,
no longer resembling a rabbit due to erosion

appears as a cloud
he night sky.

ASTRONOMY: Bryce Canyon National Park is well known for its stargazing, since up to three times as many stars can be seen here as in rural areas. The park has night sky programs many days of the week during the warm season. A variety of one-hour multimedia astronomy programs are presented, followed by up to two hours of telescope viewing. The sun may also be viewed through a special telescope during the day.

The Milky Way, the galaxy in which we live, is a spiral galaxy containing over 200 billion stars and stretching 100,000 light years in diameter. It is one of an estimated 500 billion galaxies in the universe. The size is hard to comprehend, but consider this. Using the "bucket" of the Big Dipper as a "lens" to the night sky, that area alone contains over a billion galaxies, none of which are visible to the naked eye. Amazing! An evolutionary interpretation of the universe is that it happened as a result of a "big bang." But a biblical view says the stars were spoken into existence on the fourth day of creation.

Table Cliffs
Plateau

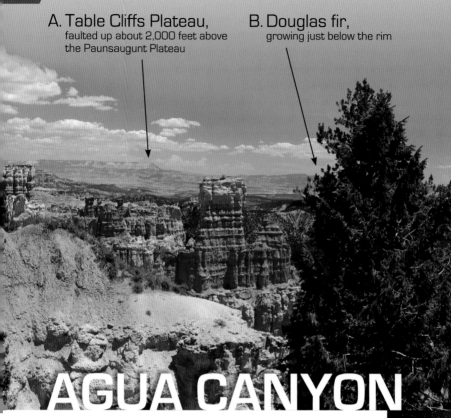

A. Table Cliffs Plateau,
faulted up about 2,000 feet above
the Paunsaugunt Plateau

B. Douglas fir,
growing just below the rim

AGUA CANYON

GEOLOGY: Agua Canyon provides a narrower view than most of Bryce's overlooks. The two hoodoos showcased here, The Hunter (C) and The Rabbit (D), provide a clue into their formation. Look closely at the top of each of the hoodoos and notice the relatively flat layer, which is of a slightly different texture. That thin layer is dolomite, which is more resistant to erosion.

The contact between the White and Pink Members of the Claron Formation in the Table Cliffs Plateau (A and below) is level and smooth like the other huge rock layers in this region whose boundaries are generally flat and even. If thousands of years had elapsed between the deposit of one layer and the next, then there should be a noticeable amount of erosion and unevenness between the layers, but there is not. This evidence suggests that the whole sequence accumulated quickly, one layer after another (see page 146).

The Milky Way
across t

Agua Canyon overlook is the viewpoint for two famous hoodoos, The Hunter and The Rabbit. In a relatively short amount of time, erosion has changed the shape of many hoodoos so they no longer resemble that for which they were originally named. The National Park Service no longer names hoodoos and even drops some of the lesser known ones from new publications.

FAST FACTS

> Bryce Canyon has some of the best visibility in the United States for star gazing.

> Sedimentary layers commonly cover huge areas and show evidence of rapid deposition, as expected from the Flood.

> Mountain chickadees, only five inches tall, never seem to stay still as they search for food hiding in the bark of trees.

> The lack of erosion between layers is evidence against the concept of millions of years.

The silhouettes of the hoodoos often form shapes with which the imagination can play, like The Hunter on the left and The Rabbit on the right. So who or what do you see in the hoodoos?

Question: Why do hoodoos not always look like their name?

E. Fin,
which erodes into a hoodoo
and then eventually collapes

F. Color variations,
caused by oxidation of iron and
other chemicals

ELEVATION 8,800 feet

*The heavens are telling of the glory of God; and their
expanse is declaring the work of His hands.*
Psalm 19:1

A. Fins and hoodoos
of the Pink Cliffs, formed by
variable erosion of soft and hard
rock layers

B. Grey Cliffs,
eroded into hills and valleys

YON

GEOLOGY: To the southeast, the White Cliffs (C) can be seen, with deep, north-south canyons carved on the edge. Similar canyons are also carved over 2,000 feet deep on the edge of the White Cliffs in Zion National Park.

The top of the White Cliffs represents a rolling erosion surface. Such erosion surfaces could have been carved rapidly by Flood water, while the deep narrow canyons at the edge of the White Cliffs likely were carved late in the Flood by more channelized receding Flood water flowing south out of Paria Valley (see page 136).

The formation that makes up the Grey Cliffs (B) is well-eroded into hills and valleys just to the east of Ponderosa Canyon. This erosion likely occurred during the Genesis Flood, but ample post-Flood erosion has also occurred.

ECOLOGY: Ponderosa pines do not grow at this overlook, but rather grow in the canyons below, where some of the 200 to 300 year old pines have trunks four to five feet across. Forests like these that have never been logged are called "virgin" or "old growth" forests. The evergreen trees along the rim are blue spruce, Douglas fir, limber pine, and white fir.

One of the reasons for the great diversity of plants and animals in Bryce Canyon is that Bryce is on the boundary of three major geographical areas. These are: 1) the Colorado Plateau, which stretches north, south, and eastward to the Rocky Mountains; 2) the Rocky Mountains themselves; and 3) the Basin and Range area to the west.

Similarly, biologists have divided Bryce Canyon into three main ecozones defined by their dominant plants, typically one or two types of trees or shrubs that characterize a zone. These ecozones are: 1) the Canadian ecozone with spruce-fir forests above 8,500 feet; 2) the transition ecozone with open ponderosa pine forests mainly at 7,500 to 8,500 feet; and 3) the pinyon-juniper ecozone at 6,500 to 7,500 feet dominated by pinyon pine, Utah juniper, gamble oak, and sage, which are mainly below the rim.

Historically, lightning strikes start fires in the pine forests every ten to twenty years (see page 159). The forests are at their healthiest when periodic, low intensity fires burn slowly and thin out the small

PONDEROSA CAN

trees and brush. Even if its green needles are scorched, the large ponderosa pines will recover; only the hottest crown-fires will kill them. As long as the inner bark that transports sugars isn't burned, the trees will survive.

PONDEROSA CANYON

Ponderosa Canyon is so named because of giant ponderosa pines in the canyon below. However, the rim of this overlook is surrounded by mostly Douglas fir and white fir. In the broad Paria Valley to the east, you can see the top three stairs of the Grand Staircase: the White Cliffs to the southeast, the eroded Grey Cliffs just below the overlook, and the Pink Cliffs to the northeast on the Table Cliffs Plateau.

N

FAST FACTS

> The lack of vegetation on the slopes below causes increased erosion of the canyon.
> The evidence suggests that rapid erosion carved the Grand Staircase and canyons in the White Cliffs to the southeast.
> The inner bark of ponderosa pines is very high in antioxidants.
> Some of the trees in the canyon below are over 150 feet tall and 5 feet in diameter.

Douglas firs can reach over 100 feet in height with short flat needles wrapped around the branch.

White firs grow up to 50 feet tall with short flat, silver-blue needles pointed upwards from the branch.

Limber pines are usually found along the rim and are less than 30 feet in height with 2 to 3 inch needles growing in groups of 5 wrapping around the branches.

Question: How many ecozones does Bryce Canyon have?

C. White Cliffs,
showing deep canyon erosion at its edge, similar to Zion National Park

D. Rainbow Point,
the southern-most overlook in Bryce

ELEVATION 8,904 feet

Ponderosa pines can grow to over 120 feet tall in the canyons below and have long needles that develop in groups of three, wrapping around the branches. The burnt-orange bark grows in a distinctive jigsaw pattern. Native Americans often ate pine nuts from the cones raw or made them into bread. They also used the pitch from the trees to make torches or as waterproofing for baskets.

A. White fir, most widespread of western fir trees

B. Aquarius Plateau, capped by 2,000 feet of volcanic rock

NYON

The blue flax has white to deep blue flowers with five petals and stands six to 31 inches tall. It is found on well-drained prairies and meadows. Native Americans ate the seeds for their flavor and nutrients. A tea was made from the stems and leaves and was used to treat various medical conditions such as eye infections, stomach disorders, and swellings. Interestingly, livestock feeding on blue flax become drowsy. Blue flax was also used by many Native American groups to make strong rope. Today flax is grown for its linen fiber and linseed oil.

Blue Flax

GEOLOGY: Bryce Canyon is not really a canyon but a series of breaks or escarpments eroding back into the Paunsaugunt Plateau. Most of this escarpment probably was formed by receding Flood water during the formation of the Paria Valley. However, Bryce's intricate hoodoo and spire formations most likely began after the Flood during the heavy Ice Age precipitation.[13]

ECOLOGY: It's interesting that limber pines, like bristlecone pines (see page 160), are found growing mostly along the rim. Apparently they can stand the harshness of the rim climate, but are crowded out by the other trees back in the forest. The limber pines standing on their roots below the rim are a testimony that these cliffs are still eroding rapidly today (see page 80).

Some people believe that a woodpecker is hardly different than most birds. But the northern flicker, like other woodpeckers, is considerably different. They hammer at the tree with incredible force — as much as 1,000 times the force of gravity! The northern flickers have several features that enable them to drill holes in trees and catch the fleeing grubs. Stored

Northern red-shafted flicker

BLACK BIRCH CA

around the back of his skull (see diagram right), the flicker's amazing tongue is as long as its body.

Yet such a special tongue would be useless without the flicker's extra neck muscles, his extra-strong beak, and the shock absorbers between the beak and skull. It has special toes to grip the tree, stiff tail feathers used for support, and eyelids that close at contact to protect its eyes. Did all these features evolve at the same time by serendipitous chance? Or is it more sensible to believe that here again is another example of the Creator's ingenious design.

Northern flicker's tongue shown in red

Black Birch Canyon is a narrow canyon and one of the smaller over-looks. It is an interesting place to study some of the shapes of the hoodoos and consider their formation. Note that some come almost to a point, while others have very flat tops. Look down in the canyon at the two pointed hoodoos standing like little twin brothers. Notice also that they both have a small "cap," which delays erosion of the softer underlying layers.

FAST FACTS

> Bryce Canyon is actually an eroded cliff or escarpment, rather than a canyon or system of canyons like Zion National Park.
> Limber pines standing on their roots demonstrate rapid present day erosion.
> The tongue of the flicker is as long as its body!
> Heavy Ice Age precipitation is likely responsible for starting hoodoo formations in Bryce Canyon.

Red squirrels harvest a mushroom-like fungus attached to Douglas fir roots. In exchange for living on the sugars in the fir roots, the hair-like filaments of the fungus greatly extend the Douglas fir's ability to absorb water and nutrients. The squirrels, fungus, and fir trees all live in a wonderful cooperative interdependence.

Question: How does the fungus growing on Douglas firs benefit them?

C. Hoodoos,
caused by variable, but rapid erosioin

D. Table Cliffs Plateau,
a flat erosional remnant of the once extensive Claron Formation

ELEVATION 8,750 feet

The largest mammal in Bryce is the Rocky Mountain elk, though they are not seen as often as mule deer (pictured) and pronghorn antelope. During the winter, these animals descend from the snow-clad plateau into the valleys and foothills below.

C. Aquarius Plateau,
volcanic rocks on top of the northern Table Cliffs Plateau

D. Eroded valley,
eroded in the formation that makes up the Grey Cliffs farther south

s Plateau,
he northeast

a significant problem for the millions of years hypothesis: the hard volcanic rock was eroded away while the underlying soft Claron Formation hardly eroded at all.

The only way we can explain the lack of erosion of the top of the Claron Formation is if the volcanic rock was eroded quickly, with very little erosion happening afterwards (see diagram bottom left). Rapid erosion during a catastrophic event best explains this dilemma.

ECOLOGY/HIKING: The Bristlecone Pine Trail starts at Rainbow Point and leads one mile to a small group of bristlecone pines found on a point exposed to the elements. Unlike other pines, bristlecone needles remain on their branch ends for many years, which give them the appearance of a bottlebrush or foxtail. The oldest tree here is estimated to be 1,600 years old.

Ancient dead bristlecone pines found in California are claimed to be almost 10,000 years old. These old ages are based on matching the pattern of wide and narrow growth rings of many dead trees. Yet, these dates are a problem because they pre-date Noah's Flood. However, heavy Ice Age precipitation, combined with little seasonal contrast during the early and mid part of the Ice Age,[13] can account for extra rings grown in one year (see page 160).

A. Canyon floor forest,
mostly made up of ponderosa pine
and white fir

B. Volcanic rocks,
capping the northern Paunsaugunt
Plateau 25 miles to the north

RAINBOW POINT

Table Clif
20 miles to

GEOLOGY: From Rainbow Point, the dark volcanic rocks capping the northern Paunsaugunt Plateau (B) and the thick volcanic rocks of the Aquarius Plateau (C) to the northeast show up well. It is likely these volcanic rocks, 2,000 feet thick, once extended south and covered the Table Cliffs (right) and the southern Paunsaugunt Plateaus.

Evolutionary scientists believe the Grand Staircase eroded very slowly over many millions of years. However, the erosion of the volcanic rocks from the top of the Paunsaugunt and Table Cliffs Plateaus demonstrates the Grand Staircase eroded rapidly.[17] This is

Aquarius Plateau

**Hard Lava
(Mostly Basalt)**

Rapid lateral erosion (cliff retreat)
proceeded while little erosion of the
plateau surface occured.

Table Cliffs Plateau

Soft Claron Formation Strata

Pink Cliffs

Rainbow Point presents an amazing view of the canyons below and is the end of both the park road and the southern Paunsaugunt Plateau. You may have noticed the increase of more than a thousand feet in elevation from the overlooks in the northern part of the park. This is because the top of the Paunsaugunt Plateau and all formations of the Grand Staircase dip downward slightly toward the north to northeast (see page 140).

FAST FACTS

> Volcanic rocks cap the top of the northern Paunsaugunt Plateau to the north and the Aquarius Plateau to the northeast.
> Little erosion occurred on the soft, exposed Claron Formation after the hard volcanic rocks were rapidly eroded.
> This region was originally inhabited as much as 4,000 years ago by people migrating from the north.
> Bristlecone pines are thought to be one of the oldest living plants on earth.

This [rainbow] is the sign of the covenant which I am making between Me and you and every living creature that is with you, for all successive generations; I set My bow in the cloud, and it shall be for a sign of a covenant between Me and the earth.
Genesis 9:12–13

Rainbow over
Bryce Canyon

Question: What evidences are there of rapid erosion of sedimentary rock?

E. Grey Cliffs,
the fourth from the bottom
cliff in the Grand Staircase

F. Erosional canyons,
carved into the top of the
White Cliffs

ELEVATION 9,115 feet

*A bristlecone pine named "Methuselah" is thought to be
the oldest living tree alive today. It lives in an undisclosed
location in the White Mountain range of
southeastern California. Named after the
longest-living person in the Bible, it is
supposedly over 4,600 years old. This
would have been close to the time of
Noah's Flood.*

C. Riggs Spring Loop Trail,
an 8.5-mile hike descending 2,200 feet
providing access to 3 back-country campsites

D. White Cliffs,
third step above the
Kaibab Plateau

anita leaves all
ting upward

rapid. This volcanic mountain, 10,388 feet above sea level and 6,000 feet above the surrounding land, was once covered by sedimentary rock layers. If erosion was slow over millions of years, it should have eroded away with the rest of the material (see page 146). Such tall erosional remnants are consistent with the erosion during a global flood.[18]

The rapid erosion of the Grand Staircase left behind two main erosion surfaces (see page 148). The first is on the Kaibab Plateau (E). The second (B) is the wide area on top of the White Cliffs (D). This surface corresponds to the Kolob Terrace in Zion from which the canyons of Zion were carved.

ECOLOGY: The evergreen manzanita shrub found growing along the rim has smooth reddish bark and round leathery leaves. On hot days, it has the ability to align its leaf edges toward the sun. Notice the vertically aligned leaves, all pointing up, in the photo above. How could a plant evolve the complex biochemical systems required to align its leaves in order to reduce the sun's effects?

Red bark of the manzanita

A. Navajo Mountain,
82 miles away, intruded into sedimentary rocks that subsequently eroded away (image enhanced)

B. Rolling erosion surface
on top of the Navajo Sandstone whos southern edge forms the White Cliffs

YOVIMPA POINT

Manz poi

GEOLOGY: The Grand Staircase can be seen quite well at Yovimpa Point. To understand its origin, visualize the Pink Cliffs (F) and all the sedimentary rocks extending southward as far as the eye can see. There would be 6,000 to 10,000 feet of sedimentary rocks covering the Kaibab Plateau (E) extending beyond the Grand Canyon. This material was then eroded away by a massive amount of water in a process called sheet erosion, forming the five "steps" of the Grand Staircase. The amount of rock removed was 100 times the rock removed in forming Grand Canyon! The evidence suggests the canyons were carved by channelized erosion after the sheet erosion phase of the retreating Flood water (see page 141).

At the bottom of the Grand Staircase, just above the Kaibab Plateau (E), are the tall Vermillion Cliffs (pictured far right). The next step to the north is the White Cliffs (D). Just below Yovimpa Point are the Grey Cliffs, which have mostly been eroded into ridges and valleys. The Pink Cliffs (F) make up the top step of the Grand Staircase.

Domed-shaped Navajo Mountain (A), seen on a clear day 82 miles away, is an indication the erosion of the Grand Staircase was

Yovimpa Point is a short hike from Rainbow Point and presents a different aspect of the Grand Staircase. The view south shows four of the five cliffs of the Grand Staircase; the first step, the Chocolate Cliffs cannot be seen. On a clear day, you can even see the Kaibab Plateau well to the south and Navajo Mountain on the horizon to the east-southeast. The formation making up the Grey Cliffs, fourth from the bottom, shows up but has been eroded into ridges and valleys just below Yovimpa Point.

FAST FACTS

> 10,000 feet of sedimentary rock eroded away to form the Grand Staircase.
> Bryce Canyon was set aside as a national park in 1928.
> The erosion surface on top of the White Cliffs is rough and rolling.
> The manzanita shrub is able to align its leaves with their edges toward the sun to reduce evaporation.

Great Basin rattlesnake

The Great Basin rattlesnake is gray to light brown in color and reaches up to five feet in length. A subspecies of the western rattlesnake, they range from northwest Arizona to southeast Oregon and eastward into southern Idaho. Often spending the winter in a common den, the female gives birth in the fall to as many as 20 young, which may live for almost 20 years. Diagnostically identifiable by their large triangular heads, the Great Basin rattlesnake is sometimes mistaken for a Great Basin gopher snake, which will "rattle" its tail in the brush to mimic the sound of the rattlesnake. How did it "learn" to do that?

Question: What caused the rolling erosion surface on top of the White Cliffs?

E. Kaibab Plateau,
80 miles away, below the
"steps" of the Grand Staircase

F. Pink Cliffs,
top step of the Grand Staircase
into which Bryce Canyon National
Park is carved

ELEVATION 9,115 feet

HIKING: The Riggs Spring Loop Trail begins at Yovimpa Point for a rugged nine-mile loop the long way to Rainbow Point. The Under-the-Rim Trail starts halfway between Yovimpa Point and Rainbow point and winds 23 miles through the backcountry to Bryce Point.

The Vermilion Cliffs (above) are named for the brilliant red color caused by iron oxide (rust) in the rock. This cliff is the second cliff up from the bottom of the Grand Staircase. It can be traced from north to east of the Grand Canyon and is over 2,000 feet tall in places.

C. Sinking Ship,
7,405 feet above sea level, tilted due to faulting

D. Fairyland Canyon,
likely eroded after the Flood

HIKING: The Rim Trail connects Fairyland Point with Sunrise Point to the south. The Fairyland Trail descends into Fairyland Canyon and loops around to Sunrise Point where the Rim Trail can be taken back to Fairyland Point. A few miles down the trail, one can visit a hoodoo graveyard in Campbell Canyon. Hoodoos which once stood tall are now just mounds of dirt and rocks. The loop is about eight miles long and strenuous. As with all hikes in the park, remember to bring plenty of water.

This point is not called "Fairyland" for nothing. The hoodoos below the point are spectacular, and you can see them at eye level while hiking down the Fairyland Trail.

ECOLOGY: The side-blotched lizard, with dark blotches on its side behind the forelegs, is one of the most abundant and commonly observed lizards in the southwestern deserts. Normally reaching about six inches, including the tail, the male is slightly larger than the female and often has bright colors on his throat. They are often confused with the western fence lizard, which is slightly bigger and has spiny scales.

dee, a common
ce Canyon

Side-blotched lizard

A. Late summer monsoons,
can quickly dump several inches of rain
from July through September

B. Table Cliffs Plateau
part of the upper step, the Pir
Cliffs, of the Grand Staircase

FAIRYLAND POINT

Mountain Chicka
visitor to Br

GEOLOGY: Fairyland Point offers a commanding view of the Table Cliffs (B) and Aquarius Plateaus to the northeast. A fault in the Paria Valley to the east has uplifted these plateaus, raising them 2,000 feet higher than Bryce. The Aquarius Plateau is the top of 2,000 feet of volcanic rock that used to cover the Table Cliffs Plateau. Mysteriously, as the hard volcanic rock eroded northward, the soft underlying Claron Formation was not eroded. From a creationist's perspective, this can best be explained by rapid erosion during the runoff of the Flood water (see page 123).

Along the rim at Fairyland Point are numerous blocks of rock that contain small water-rounded rocks cemented together. This is called conglomerate, and patches of it are found throughout the Claron Formation, especially here, on Boat Mesa (F), and at Bryce Point. The small, hard rocks do not come from the Bryce Canyon area, but appear to have been transported by water from mountain ranges in eastern Nevada. Such long distance transport of resistant rocks is consistent with a global flood.[18]

Fairyland Point is actually accessed from outside the park, about one mile to the north. From Highway 63, turn east and drive about one mile to the overlook, which offers a good view of the Table Cliffs Plateau to the northeast. Looking farther north, the dark edge of the Aquarius Plateau can also be seen standing 11,300 feet above sea level and is the highest plateau in the United States.

FAST FACTS

> Volcanic rock on the Table Cliffs Plateau rapidly eroded toward the north, exposing the soft Claron Formation.
> Mysteriously, the soft Claron Formation hardly eroded at all.
> Rounded rocks in the conglomerate at Fairyland Point likely came from eastern Nevada 200 miles away.
> Side-blotched lizards are most commonly seen below the rim on dry slopes with sparce vegetation.

The Sinking Ship is a large fault block of Claron Formation rock. It was tilted downward toward the northwest by movement on the Paunsaugunt Fault, just east of Sinking Ship.

Question: What could have transported and rounded the small rocks in the conglomerate along the trail?

E. Lone hoodoos
standing high above most
others in the area

F. Boat Mesa,
capped by conglomerate with
rounded rocks transported from
far to the west

ELEVATION 7,758 feet

Hummingbirds are a delight to watch as they hover over flowers sipping
nectar. Their feathers are not colored but rather refract the light, like
a crystal or rainbow, to create the perception of color. Hummingbirds
can beat their wings up to 50 times a second, fly
backwards and sideways, take off vertically, and
fly on their backs. This amazing bird shows
the glory of our Creator: ". . .Your works are
wonderful, I know that full well" [Ps. 139:14].

3.5 inch rufous hummingbird
sipping from a thistle flower

Bryce   Zion

Why is geology often in the center of the creation/ evolution debate? It is because geology is the foundation of evolution. The supposed long ages and slow processes of geology seem to make the evolutionary story plausible even though the fossils and biology strongly favor creation.[19] Evolution, to be feasible at all, requires billions of years for life to evolve and for the thick sedimentary rock layers to form. However, there is worldwide evidence that the majority of the fossil-bearing sedimentary rock formations seen today are the result of a recent and catastrophic global flood. This provides significant support for the flood of Noah's day as described in Genesis.

The book of Genesis is foundational to the entire Bible, for in Genesis many great biblical themes find their origin. If the accounts in Genesis are not true, why should the rest of the Bible be considered as truth?[20] But, if the Flood account in Genesis is true, man is then confronted with a Creator, a Creator to whom he is accountable, a Creator who tells us in Psalm 19:1, *"The heavens are telling of the glory of God; and their expanse is declaring the work of His hands."*

Everyone has a worldview, whether they realize it or not. Your worldview includes those preconceived ideas and biases through which you view the world. As you look at the Grand Staircase, ask yourself, "What do I see here, and of what importance is it to me?" As you consider this question, also consider your worldview and how it affects your analysis of the significance of this amazing place.

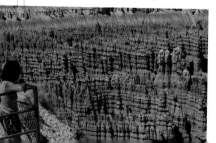

Bryce Canyon's Inspiration Point

WORLDVIEW ASSUMPTIONS

Before we can understand how the Grand Staircase, including Zion and Bryce Canyons, was formed, we first need to understand the underlying worldview assumptions made by those providing the information. Worldview assumptions are primarily based on our beliefs about the origin of all things. Since we all look at the same physical data, the same rocks, it is those assumptions which affect the outcome of any analysis. So as a starting point, we need an understanding of the process of "science."

True scientific knowledge advances by careful observation and testing that can be repeated by others, a process referred to as "operational science." Thus, scientific theories and facts are established by repeated similar results. According to this definition, neither creation nor evolution are true science. Both could be called "historical

> True scientific knowledge advances by careful observation and testing that can be repeated by others...

science," but really are just plain history. They are scientific ideas or historical models about events that happened in the unobservable, untestable past. Nevertheless, the common tools of operational science, observation, and testing should be used as the basis for determining which historical model (evolution or creation) fits more closely with what is observed in the world around us.

Interpretations of the Grand Staircase, specifically its geology, can be divided into two categories or worldviews — the evolutionary/uniformitarian model and the creation/Flood model. These are two completely different ways of interpreting the world. Either the rock layers of the Grand Staircase represent millions of years of slow processes, or they are a monument to the global Flood of Noah's day. Both models can't be correct.

Take the simple illustration shown below of how two scientists differently interpret the same data based on their worldviews. Note that both scientists start with preconceived ideas (their worldview) as they interpret the data. These ideas drastically affect their analysis and their conclusions.

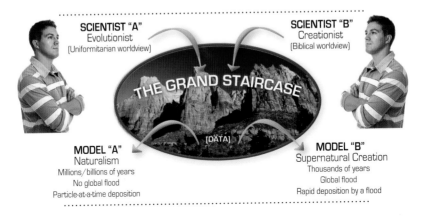

SCIENTIST "A"
Evolutionist
[Uniformitarian worldview]

SCIENTIST "B"
Creationist
[Biblical worldview]

THE GRAND STAIRCASE

[DATA]

MODEL "A"
Naturalism
Millions/billions of years
No global flood
Particle-at-a-time deposition

MODEL "B"
Supernatural Creation
Thousands of years
Global flood
Rapid deposition by a flood

Thus, as you analyze the interpretative literature and park signs about the Grand Staircase, consider the worldview presented in the material. Since man did not witness the formation of the Grand Staircase, it is his worldview that influences analysis and therefore his conclusion. This True North Guide interprets the data in light of a biblical worldview (Scientist "B").

Creationists do not pretend to have all the answers. What they do contend is that their interpretation, based on the Bible's account of history as well as the scientific evidence, provides a more plausible and cohesive explanation of earth history than the evolutionary interpretation. Creationists believe their interpretation is more consistent with the data seen on earth and with the information given us by the One who was present when the Grand Staircase was formed — God.

HOW WAS THE GRAND STAIRCASE FORMED?

When viewing Zion Canyon's amazing, sheer cliffs or Bryce Canyon's vast display of intricate formations, one cannot help but wonder how these features formed. Since they are part of the Grand Staircase, the question of how the Grand Staircase formed must first be addressed.

Before the Grand Staircase was carved, the sedimentary layers of the Colorado Plateau were deposited over tens of thousands of square miles. Almost all geologists agree that at some point after the initial deposition, erosion by water stripped many thousands of feet of sedimentary rocks off of the Colorado Plateau, including the Grand Canyon area, forming the Grand Staircase.[21] Secular geologists call this horizontal erosional event "The

Great Denudation." From an evolutionary perspective, there are a couple of difficulties with the concept of The Great Denudation; mainly we do not observe that kind of horizontal erosion happening today. Instead, what we see is mostly canyon cutting or vertical erosion. Additionally, evolutionists start with the assumption of "a little bit of water over a long period of time" which does not provide

Viewing Bryce from Sunrise Point

a mechanism for massive horizontal erosion and subsequent removal of the sediments. Creationists instead believe the majority of evidence suggests the Grand Staircase was carved by "a lot of water over a short period of time," which is consistent with the biblical Flood.

The Flood can be divided into two stages; first the flooding stage, followed by a retreating stage.[22] The flooding stage encompassed the main disaster of the global Flood up to the time the water peaked on the earth. It probably would have been during the flooding stage the layers of the Grand Staircase were deposited. During the retreating stage, the water ran off the rising continents, causing massive erosion. At first, the water would have run off as wide currents and eroded large horizontal areas in a process called sheet erosion. The Great Denudation would have taken place at this time. Then, as more and more mountains and plateaus were uplifted and exposed, the water would have been forced to flow around these obstacles in a more channelized flow. Zion Canyon, Paria Valley, and other major canyons of the Grand Staircase would have been carved as the retreating waters channelized.

The questions is, are the rock layers in the Grand Staircase a monument to slow changes of the earth's surface over millions of years as the evolutionists postulate? Or, are they a monument to rapid deposition, followed by rapid erosion during the retreating of a global flood, as creation scientists believe? We believe the evidence supports the latter.

WHY IS THE GRAND STAIRCASE IMPORTANT?

As you consider the question presented earlier, "What do I see here, and of what importance is it to me?" understand there are two areas of study that stand out: science and theology. Because if the Genesis Flood carved the Grand Staircase, including canyons like Zion and Bryce, then there are strong theological implications. We invite you to explore the evidence found in the following sections and consider their implications for you and your worldview.

For since the creation of the world His invisible attributes, His eternal power and divine nature, have been clearly seen, being understood through what has been made, so that they are without excuse.

Romans 1:20

Geology

Bryce

Zion

138

The science of geology studies rocks (in the present) and attempts to make deductions about the unobservable past based on those observations. Such deductions depend upon one's assumptions and worldview, as these bring expectations of what will be found. Some of the geologic expectations from the two worldviews are outlined in the table to the right. (This is also discussed in more detail in the *Understanding the Grand Staircase* section on page 134.)

The geological cross section on pages 142 and 143 shows the sequence of northward-tilted layers that make up the Grand Staircase and shows their relationship to the Grand Canyon layers. These formations include sedimentary layers of sandstone, limestone, and shale on top of the Kaibab Limestone that is the surface layer in the Grand Canyon region. The cross section shows the five cliffs with the top cliff covered by mostly basalt lava on the northern part (see figure on page 123). These layers, representing a thickness of over 10,000 feet, once extended far southward past the Grand Canyon.

Evolutionists explain that the earth and all life, including mankind, are the products of natural processes acting randomly on matter and energy, beginning around 4.6 billion years ago, and that everything has come about naturally without a supernatural designer. They study the rock layers based on those assumptions, while assuming the rates of erosion and deposition in the past were generally the same as the rates we observe today.

But this has been shown to be an incorrect assumption. An example of this is the glacial Lake

Windows eroded into a fin

Missoula flood that carved the Channeled Scabland in eastern Washington.[23] The 15,000 square miles of the Channeled Scabland were considered in the early 20th century to be the product of slow, gradual processes. In the mid-1920s, catastrophic processes were proposed and rejected, but 40 years later, geologists accepted the Channeled Scabland as the product of a catastrophic regional flood (not Noah's Flood).

In contrast, young-earth creationists explain that the earth, and all life, was created six to ten thousand years ago. They study the same rocks and processes studied by the evolutionists; but in addition, they study the biblical record about the origin of the earth and life. They accept as historical fact the Genesis account of a global flood (see page 17). With that as their foundation, creationists believe there is abundant geologic evidence for a global flood of catastrophic proportions, and that the geologic record and biblical account are in full agreement.

> *I establish My covenant with you; and all flesh shall never again be cut off by the water of the flood, neither shall there again be a flood to destroy the earth*
> Genesis 9:11

Are the rock layers in the Grand Staircase a monument to slow changes of the earth's surface over millions of years? Or, are they a testimony to the authority and truth of God's word and the awesome power of the Flood, as creation scientists believe?

GEOLOGIC WORLDVIEW EXPECTATION TABLE

If the geologic record were the result of processes acting over a long/short period of time, the following would be expected:	EVOLUTION (Uniformitarian) ▼	CREATION (Flood Catastrophe) ▼
	long period of time	short period of time
The extent of sedimentary rock layers would:	be small, with local origin and distribution	be massive, often almost continent-wide in distribution
The type of rock layers and erosional features would:	reflect gradual rates of deposition and erosion	reflect rapid rates of deposition and erosion
Sheer cliffs and canyon walls would:	be broken down by erosion	still exist
Erosion between layers would:	be deep and found frequently	be shallow and seldom found
Time gaps (missing ages) between layers would:	be found often, representing large gaps	be nonexistent
The geologic column with its imbedded fossils would:	show repeated uplift and erosion of the land	remain intact in many areas

FORMATION OF ROCK LAYERS IN THE GRAND STAIRCASE

Before Zion and Bryce Canyons were carved, the geologic processes of erosion, transport, and sedimentation were responsible for the formation of the rock layers of the Grand Staircase, just as most everywhere else on earth. Sediments were transported by moving water and then deposited as layers. Then they were finally "cemented" together by chemicals in the water, forming mostly sandstone and shale. Also, large volumes of dissolved calcium carbonate would have precipitated out of seawater to form layers of limestone.

The Grand Staircase is the northern edge of an eroded ridge or anticline, (a dome of rock uplifted and then eroded down, see diagram on next page). The sedimentary rocks in this anticline once extended southward over the Grand Canyon area. About 10,000 feet of sedimentary rock south of the Grand Staircase was eroded and removed over a huge area.[24] Red Butte, Cedar Mountain, and Shinumo Altar, near Grand Canyon, are approximately 1,000-foot high erosional remnants left over from this vast erosional event.[25] Navajo Mountain (see page 103) is a 6,000-foot high remnant left over from the same event.

The Grand Staircase represents a remnant of about 10,000 feet of sedimentary rock layers,[26] the northern part of which is capped by volcanic material (see diagram page 142). The layers tilt downwards toward the northeast by just a few degrees.[27] The staircase pattern is caused by erosion of less resistant rocks, leaving the more resistant layers as "steps." There are five main cliffs, all with different colors, allowing for easy identification. The Chocolate Cliffs are the lowest cliff just above the Kaibab Limestone, which forms the upper corner of the Grand Canyon. Vertically above and a little north of the Chocolate Cliffs are the 1,500-foot high Vermillion Cliffs, which consist of the reddish-colored Moenkopi and multi-colored Chinle Formations. Then comes the 2,000-foot high White Cliffs, made up of the Navajo Sandstone (see page 145). Zion National Park is carved mostly out of the White Cliffs. Next come the Grey Cliffs, which are about 3,000 feet thick, followed northward and upward by the 1,500-foot thick Pink Cliffs from which Bryce Canyon National Park is eroded. Creation scientists would agree that the sedimentary rocks of the Grand Staircase were deposited during the Genesis Flood. And most would agree the erosion that formed the Grand Staircase occurred late in the Flood as the Flood water drained off the rising continent.

Sandstone cliffs of Zion Canyon as seen from the east entrance

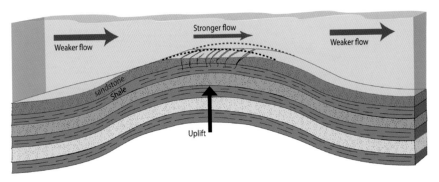

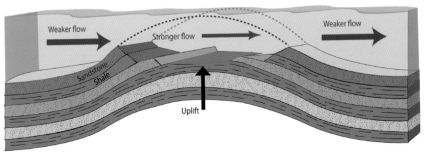

A ridge uplifting in the Flood water would be greatly eroded at the top because the rock would crack and the water speed would increase while passing over the ridge. Such an event can explain the erosion of the Grand Staircase.

EXTENSIVE EROSION

In trying to understand what we see in the Grand Staircase today, we must also consider what we do not see. The erosional remnants found here indicate the layers extended south over the whole Grand Canyon area. There are around 100,000 cubic miles of material "missing" from the area south of the Grand Staircase, including the Grand Canyon area. This is over 100 times the 900 cubic miles eroded to form the Grand Canyon. Where did all the material go? Why isn't there an enormous debris pile somewhere west, south, or east of the Grand Staircase?

Evolutionary theory suggests this material was removed by slow processes of erosion over millions of years. In contrast, creationists believe this material was removed when the Flood water retreated off the rising continents.[18] The material was rapidly swept off the continent and deposited when the currents slowed—at the edge of the continent forming some of the sedimentary rocks of the continental shelf and slope.

NINE EVIDENCES FOR NOAH'S FLOOD IN THE GRAND STAIRCASE

The following are observations and explanations of some of the geological evidence for the Genesis Flood seen in the Grand Staircase.

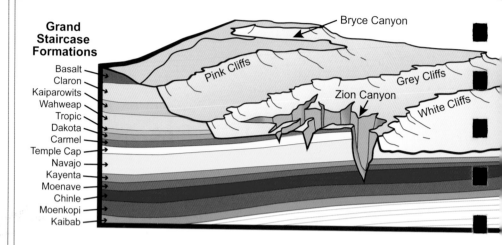

Grand Staircase Formations

Basalt
Claron
Kaiparowits
Wahweap
Tropic
Dakota
Carmel
Temple Cap
Navajo
Kayenta
Moenave
Chinle
Moenkopi
Kaibab

Bryce Canyon
Pink Cliffs
Grey Cliffs
Zion Canyon
White Cliffs

CROSS-BEDS IN THE NAVAJO SANDSTONE

The Navajo Sandstone is the thick, white layer of rock forming most of the peaks in Zion National Park. It is more than 2,000 feet thick in the park, but thins to the east and disappears in the eastern Colorado Plateau (see page 145).

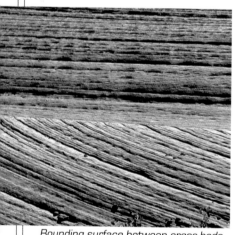

Bounding surface between cross-beds in Navajo Sandstone.

From many of the overlooks, diagonal lines called cross-beds can be seen in the sandstone. These inclined beds are formed either from underwater "sand waves" or from ancient wind-driven "sand dunes." There is a difference of opinion among geologists, even among some evolutionary geologists, regarding the environment that produced the Navajo Sandstone and its cross-beds.[12] A few hold to the marine deposit theory, but most still favor a dry sand dune origin.

Most evolutionary geologists have claimed three main evidences for what they believe are desert sand dunes of the Navajo Sandstone. These are the thickness of the beds, the high angle of the incline in the cross-beds, and the qualities of the sand grains themselves. However, all these factors are debatable.[12]

For example, studies have shown that cross-bed angles in terrestrial dunes incline from 30 to 34 degrees, whereas water-formed cross-beds incline at 25 degrees or less. But, there is a broad overlap in cross-bed dip angles between environments, making this criterion unreliable.[28]

Grand Staircase

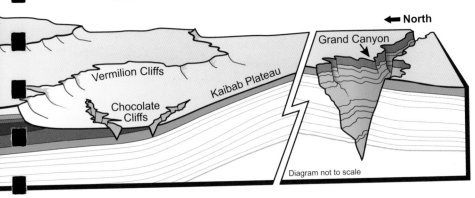

← North

Grand Canyon

Vermilion Cliffs

Kaibab Plateau

Chocolate Cliffs

Diagram not to scale

Furthermore, cross-beds are frequently separated by flat planation surfaces called bounding surfaces as pictured on page 142. Often a few dozen of these surfaces can be seen in cliffs "stacked" like pancakes, some of which can be traced for miles. There is no such process occurring today to flatten the top of a series of desert dunes without causing chaotic sand features at the boundary. But studies show that such planing can occur underwater as deposition goes from forming dunes at low current velocities to shearing the tops of the dunes at moderate velocities.

The lower and upper contacts of the Navajo Sandstone are flat, like the Coconino Sandstone in Grand Canyon.[26] Such flat boundaries are not found in deserts but would be expected during Flood deposition.

CONGLOMERATE IN THE CLARON FORMATION

Conglomerate is commonly found within the Claron Formation. Conglomerate is composed of smaller rocks that have been rounded by water and then cemented together. The smaller rocks in these conglomerates are not found in the vicinity of Bryce Canyon. The rounded rocks came from the west and are composed of types of rocks found in the mountains of Nevada, nearly 200 miles away. Such long-distance transport is rare with modern streams and rivers but would be expected in the fast flow of the Genesis Flood.

Conglomerate from the Claron Formation

NARROW, VERTICALLY WALLED CANYONS

The Grand Staircase has many narrow, vertically walled canyons, especially in Zion National Park. One example is The Narrows, a 2,000-foot deep slot canyon (see A page 48). Today narrow canyons are cut by catastrophic erosion. Many vertically walled canyons are still found in the Channeled Scabland of eastern Washington, formed after the gigantic Lake Missoula flood.[23]

Vertical walls of Zion Canyon show a lack of rockfall at their base.

It has been shown that vertical walls erode faster and lose their steepness as erosion rates are higher where slopes are steep.[29] Geologists know that vertically walled canyons soon develop a more V-shaped valley profile as the slope's steepness decreases with time.[30] So, vertical walls are a sign of a youthful canyon, as we would expect only 4,500 years after the Flood.

LACK OF ROCKFALL

As you view the canyons, one of the first things noticed are the massive cliffs, but what is not seen is the massive amount of rock that had to have collapsed and been removed when the cliffs were formed. These cliffs were not necessarily caused by a sapping process at the base of the cliff, but more by direct erosion. However, there is a conspicuous lack of rockfall debris throughout the canyons, especially in Zion National Park.

Recent and catastrophic carving of canyons is the key to this lack of material. In a millions-of-years model, there is no mechanism for the subsequent removal of the material, no matter how it was initially removed from the cliff. Evolutionary geologists believe the Virgin River eroded Zion Canyon over a few million years. If this were the case, what eroded all the deep side canyons? A catastrophic flow of water over the whole area would carve both. Such catastrophic water flow is consistent with a global flood, which would not only carve deep vertical canyons but also remove the debris.

EXTENT OF SEDIMENTARY ROCKS

The pancake-like layers of the Grand Staircase are widespread and can be traced for hundreds of miles. For example, the Shinarump Conglomerate (pictured right) is about 50 feet thick and covers more than 100,000 square miles of the Colorado Plateau.[31] This sheet-like layer has no evidence of channels in it, yet it contains rounded rocks up to six inches across that have been transported for hundreds of miles. Nowhere on earth are present processes depositing a uniform thickness of sand and gravel over such a large and generally level area. It seems more likely that this vast thin sheet was deposited by a fast current of water. Such transport and deposition would be expected during a global flood.

Shinarump Conglomerate

Likewise, the Navajo Sandstone and equivalent formations are supposed to be one of the largest wind-deposited formations in the world (see map below). It is spread over northern Arizona, Utah, Wyoming, northwest Colorado, and a small portion of eastern Nevada and southeast Idaho. Before being eroded to its current size it is believed to have covered 155,000 square miles, an area the size of California. The Navajo Sandstone thickens toward the west where it reaches a maximum of 2,200 feet in Zion National Park.

Another consideration is the origin of the material necessary to create the layers. It could not have come from the erosion of the different types of underlying layers, so the sand for the Navajo Sandstone had to originate from hundreds of miles away, just like other so-called desert deposits in the Colorado Plateau. Some evolutionary geologists believe the sand was transported from as far away as the Appalachian Mountains, based on their dates for some of the sand grains.[11]

Evolutionary explanations for the formation of the Grand Staircase layers are based on the kinds of processes we see happening around us today. But regional scale deposition implies regional scale geologic processes. The enormous horizontal extent of the layers testifies to the type of catastrophic processes expected in a global flood.

The Flood also provides a better model for understanding the source of material, the long-distance transport of it, as well as the enormous extent of the Grand Staircase layers. The evidence strongly suggests that the Navajo Sandstone was laid down in water and not in a desert environment (see page 142).

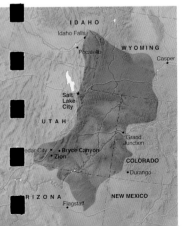

Extent of Navajo Sandstone and equivalent formations

BOUNDARIES BETWEEN LAYERS

The horizontal boundary lines between the sedimentary rock layers seen in the Grand Staircase area are called contacts. If the layers are considered to be laid down continously, one on top of the other with no gap of time between them, the boundaries are call "conformable" contacts. If there is thought to be a gap of time or erosion between the deposition of the two layers, the contact is called an "unconformity."

Within the Grand Staircase, evolutionists date the lowest/oldest layer, the Moenkopi Formation, at about 250 million years, and about 40 million years for the top/youngest Claron Formation. This represents 210 million years of multiple cycles of advancing and retreating seas that causes shallow marine and land sedimentation, including a debatable desert deposit (see page 142). Then 2,000 feet of volcanism capped the sequence 15 to 30 million years ago.

The evolutionary interpretation of most of the contacts seen in the Grand Staircase is that they represent from a few million up to 65 million years of missing time, making them unconformities.[32] For example, in the picture on the next page, the tabletop flat contact between the Navajo Sandstone and the Temple Cap Sandstone (arrow) theoretically represents a few million years of missing time and material. If all this time were real, shouldn't we see evidence of tremendous erosion at the boundaries between the layers where rivers, streams, and other erosional agents removed material? Why is there little or no sign of erosion between the layers? Why is there hardly any sign of channels, canyons, or valleys, as we see with the erosion of present-day topography (see diagram below)?

At the current rate of erosion, all of North America would be eroded to sea level in ten million years.[33] But, man is responsible for increasing (or in some areas decreasing) the current rate of erosion. Also, as the elevation of the land decreases, the erosion rate decreases. With that in mind, ten million years would be a minimum, with 40 million years likely being the maximum. Erosion is so fast that if the hundreds of millions of years of missing time is real, we should see considerable evidence of erosion between the layers seen in the walls of cliffs.

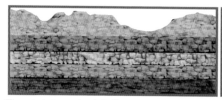 

The sketch on the left illustrates how the layers of the Grand Staircase actually appear. The sketch on the right illustrates what would be expected if significant time had elapsed between the deposition of one layer and the next.

Notice the lack of erosion between layers

Consider the two theories for the origin of the layers seen in the Grand Staircase. The assumptions and predictions of the creationist's view could not be more dissimilar to the evolutionist's view. Contrary to the millions of years seen by the evolutionists, the creationist's view is that the sedimentary layers were all deposited during a single catastrophic event of global proportions about 4,500 years ago. The expected evidence for this explanation should show continuous deposition of sediments, one layer upon another, with flat boundaries showing very little, if any, erosion between them. And that is, in fact, what is observed.

ROCK ARCHES

A natural bridge is associated with a stream or former stream, while an arch is not. Zion National Park displays dozens, if not hundreds, of arches of all shapes and sizes. Practically all these arches are small and are formed by erosion. Kolob Arch in northwest Zion National Park has a span of 287 feet and is the second largest in the world. (Landscape Arch in Arches National Park is number one at 290 feet.) Kolob Arch is different than many arches in that it is eroded from an alcove on a cliff face, now separated from the cliff by only 44 feet.

Freestanding arches, those that are narrow and supported at the ends, are more mysterious. Many of these freestanding rock arches are on ridges or the sides of a ridge or mountain. For instance, Crawford Arch (also called Bridge Mountain Arch) is a remarkably thin layer of sandstone three feet wide and 156 feet long on the side of Bridge Mountain in southeast Zion (see page 35). This arch can be seen at a distance from the Zion Human History Museum, where a display points to its location.

Kolob Arch,
world's second largest arch

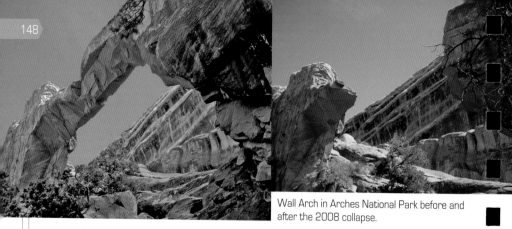

Wall Arch in Arches National Park before and after the 2008 collapse.

Evolutionary scientists believe freestanding arches are formed slowly over many tens of thousands of years by weathering and erosion. But, if it took this much time to form these arches, the arch would have collapsed long ago, since the arch itself also weathers. We observe the collapse of freestanding arches today — for instance Wall Arch in Arches National Park that collapsed in 2008 — but we do not see them being formed. It seems that freestanding arches require rapid erosion, which would occur in a flood but not by present processes of erosion.

Many of the smaller arches in Bryce Canyon could have formed in a short time by normal erosion because the rocks are a softer material. These arches are more properly called windows. Although called a natural bridge at Natural Bridge overlook in Bryce Canyon National Park (see page 106), this feature is, as the interpretive sign states, really an arch.

PLANATION AND EROSION SURFACES

According to the *Dictionary of Geological Terms*, an erosion surface is: "A land surface shaped and subdued by the action of erosion, especially by running water. The term is generally applied to a level or nearly level surface." [34] An erosion surface is regarded as a rolling surface of low relief. A planation surface is a flatter erosion surface. [35] Planation and erosion surfaces seen on the surface were cut by running water, since most of them have a flat to nearly flat topography. Many planation surfaces are capped with a thin veneer to a hundred feet of water-transported rocks (see diagram below).

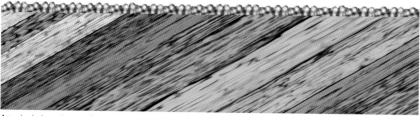

A typical planation surface on titled sedimentary rocks flattens both hard and soft layers the same, unlike erosion today. A thin layer of water-rounded rocks cap many planation surfaces.

Planation surfaces are not forming today, except rarely when a river floods and erodes the rock along the bank. But such river erosion is very small scale. Yet planation surfaces are found worldwide and sometimes cover over a thousand square miles. Erosion by retreating Flood water can explain them.

Planation and erosion surfaces are common in the Grand Staircase area. The first erosion surface is the area around the Grand Canyon on top of the Kaibab Limestone, the very bottom of the Grand Staircase. The next significant erosion surface in the Grand Staircase is Kolob Terrace. This erosion surface is moderately dissected with a few hills rising above the level of the surface (see page 71). These hills are erosional remnants and are called monadnocks in geological jargon. The Kolob Terrace is imprinted on the top of the White Cliffs.

The canyons of Zion National Park have been cut vertically down a few thousand feet through the Kolob Terrace after the erosion surface was formed. This is similar to the formation of Grand Canyon after the planation surface eroded on top of the Grand Canyon area. Deep canyons were also cut on the Kolob erosion surface south of the Paria Valley along the edge of the White Cliffs (see page 125).

The bounding surfaces (sometimes called truncation surfaces) in the Navajo cross-bedded sandstone pictured below are an example of a planation-like surface, but within the formation. These bounding surfaces indicate that the tops of cross-bedded sand dunes were rapidly planed off by periodically faster currents. When the current slowed down, the deposition of cross-bedded dunes resumed. These bounding surfaces would be just as expected for deposition during a global flood.

Bounding surfaces
(note car for scale)

UNDERFIT STREAMS

Valleys and canyons all over the earth provide evidence that they carried much more water than the stream or river that flows in them today. The stream or river in the valley is considered underfit. This relationship applies to both nonglaciated areas and glaciated areas, where meltwater from ice would have provided much more water than today.

The amount of water that once flowed in these drainages is related to the size of the meanders of the valley or canyon compared to the stream today.

The meander size provides an estimate of the original flow volume. It has been estimated that when the valley or canyon formed, based on the large size of the meanders, they contained 20 to 50 times the volume of water as the current stream.[36] Such a large volume of water is shown for Zion Canyon with its much larger meanders than the current Virgin River (see page 48). Channeling of Flood water late in the Flood would easily explain the evidence for much more water carving deep canyons and valleys.

The stream here is "underfit," being too small to have carved this canyon.

DATING METHODS

Methods used to date rocks fall into several categories. Carbon dating, explained below, is often thought to be used to date rocks. But carbon dating is most useful in dating organic matter (i.e., matter that was once living) that has not been fossilized. Radiometric dating is used by evolutionary geologists to date metamorphic and igneous rocks, while fossils are used to date sedimentary rock layers. But do they provide meaningful results?

CARBON DATING

Carbon-14, which is absorbed by living matter at a known rate, decays into nitrogen-14 at a known rate. Geologists test levels of carbon-14 remaining in organic material, such as coal, to determine its age.

Current research in carbon-14 dating, agreed by most scientists to be only accurate up to a maximum of 70,000 years, is providing further evidence of a young earth. Carbon-14 has been found in coal samples, some from the Colorado Plateau, thought to be up to 350 million years old. That is 5,000 times carbon-14's maximum life! Additionally, coal samples from different layers or "ages" in the geologic column have almost identical amounts of carbon-14, suggesting that they are really all the same age.[37]

RADIOMETRIC DATING

The measurement of radioactive elements as they decay from a parent element to its daughter element, which occurs at what is thought to be at a known rate, is a process called radioactive decay.[37] Evolutionists use this decay process to determine the ages of igneous and metamorphic rocks. There are several elements used in this type of dating method. For example, potassium is known to decay into argon, and rubidium into strontium.

Three basic assumptions are made in radiometric dating: 1) the rock starts out with known quantities, or ratios, of both the parent and daughter elements, 2) nuclear decay rates never change, and 3) the decay process takes place in a "closed" system that does not allow for elements to leach into or out of the rock. The erroneous and vastly different ages frequently found by geologists indicate that all of these assumptions are subject to failure. For example, the analyses of samples taken only one yard apart from the same formation, at the same time, and using the same dating method has produced dates over one billion years apart.[37] If radiometric dating were dependable, how could two samples of the same rock produce such dramatically different ages?

SEDIMENTARY DATING

Sedimentary layers are dated based on their relative position in the rock record sequence and by the fossils found in them. Specific fossils, called "index fossils," are thought to indicate the age of the formation. But the age of a fossil is determined by the layer in which it is found.

If a paleontologist is dating fossils, like the petrified wood pictured below, they would first look to see how old the layer is in which they found the fossil to determine its age. The way they do that is by looking at the fossils found in that layer. So, they date the fossils by the layer in which they are found, but they date the layer by the fossils found in it. Do you see a bit of circular reasoning in this process?

Petrified
wood

SECTION NINE

Fossils

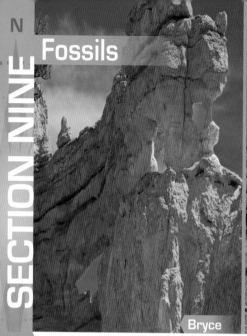

Bryce   Zion

Fossils are the evidence of past life and include preserved remains of living organisms, imprints of plants and animals, and preserved animal trails or burrows found in sedimentary rock layers of the earth's crust. Some preserved remains, such as wood and animal bones, are commonly found completely replaced by minerals and are sometimes found by the thousands buried in huge fossil graveyards.

FOSSIL RECORD SUPPORT OF THE CREATION MODEL

The creation model and the evolution model for fossilization must be evaluated by careful study and interpretation of the evidence in the rock layers and their fossils. The conditions necessary to make a fossil are uncommon today. First, the organism must be buried quickly to protect it from decay and scavengers. Rapid burial happens today only on rare occasions. Second, minerals must be absorbed rapidly within the hard parts before the organism decays in the ground. Today, there are rarely enough minerals within groundwater to fossilize organisms. Noah's Flood would have been ideal for forming fossils rapidly, as the Flood would result in quick, deep burial and extreme pressures, causing the mineral-charged water to flow into the organism. A global flood best explains the broad regional extent of fossil-bearing sedimentary rock layers in Zion and Bryce Canyon National Parks, as well as worldwide.

Instead of "proving" evolution, the fossil record actually supports the creation model and the record of the biblical Flood. The table on the next page compares what the two different worldviews might expect to find as the fossil record is examined.

Fish fossil

LACK OF TRANSITIONAL FOSSILS

The general sequence in the fossil record is often used as a "proof" of evolution over millions of years. But if plants and animals evolved slowly over time, why don't we find the intermediate or transitional forms of life one would expect in the fossil record? What we find instead are complex, fully developed, fully functional creatures right from their very first appearance, which is as expected for the special creation of all plant and animal kinds within a matter of just a few days.

Evolutionary scientists admit that all major groups (phyla) of living things appeared abruptly on earth at the same time, about 540 million years ago, in what they call the Cambrian Explosion. They have no explanation for how fossils of all these groups could have suddenly appeared. This was a quantum leap for evolution from the single-celled life-forms to the broad array of fossils of fully developed, complex organisms with no linking transitional fossils found.

Charles Darwin, author of *The Origin of Species*, said, "Why then is not every geological formation and every stratum [layer] full of such intermediate links? Geology assuredly does not reveal any such finely-graduated organic chain: and this, perhaps, is the most obvious and serious objection which can be urged against the theory. The explanation lies, as I believe, in the extreme imperfection of the geological record."[38] Darwin thought the problem was that not enough of the fossil record had been explored and that the transitional forms would be found. But since then, literally millions of fossils from across

FOSSIL RECORD WORLDVIEW EXPECTATION TABLE		
If the fossil record were the result of processes acting over a long period of time, versus the result of a global flood, the following would be expected:	**EVOLUTION** (Uniformitarian) ▼ long period of time	**CREATION** (Flood Catastrophe) ▼ global flood
The fossil record would show:	multiple thousands of transitional forms	distinct, complex, fully functional forms
Fossils of plants and animals normally living together would:	be commonly found together	be rare due to the sorting power of a global flood
The total number of major life-forms (phyla, classes, and orders) would have:	increased upwards as life-forms evolve	decreased from the total created number because of extinction
Location of simple versus complex life found in the fossil record would:	increase upwards in complexity from simple to complex	be complex throughout the fossil record
The order in which fossils are found in the rock layers would:	show stages of an evolutionary sequence	show a rapid burial sequence of created complex creatures

the globe have been documented, and still no fossils have been found that clearly represent transitional life forms.

FOSSIL COMMUNITIES UNCOMMON

If organisms were buried by local floods or on the bottom of seas, as proposed by the evolutionary model, you would expect to find fossil communities buried together. For example, the burial of an intertidal habitat, such as the coral reef shown above, would bury a variety of creatures such as starfish, snails, anemones, barnacles, sea urchins, and marine plants that lived together. But we do not see fossil communities in the rock layers. Sea creatures from communities appear to have been moved and buried by sediment-carrying currents, rather than buried slowly in a calm placid sea as the evolutionists suggest. Currents moving over vast areas would sort and separate creatures from their communities, as expected in a catastrophic flood of global proportions. Also, community beds of fossil clams all in their "natural" position should be common, but it is significant that they are not. Instead, fossil clams are found oriented in every direction with their shells commonly still closed, indicating they were transported and still alive when buried.

FOSSIL EVIDENCES SEEN IN THE PARKS

Evolutionists believe the fossil record shows the continuing evolution of living things over millions of years. For creationists, the fact that fossils of the same type of organism can sometimes be found in multiple rock layers, for instance in the Grand Canyon, is best explained by a catastrophic global flood.

The fossils found in Zion and Bryce Canyon National Parks, including plants, fish, amphibians, reptiles, and small mammals, provide insight into processes and events during the catastrophic global flood that buried and fossilized so many living things. The well-preserved nature of many fossils found in the parks is also testimony for rapid burial.

VERTEBRATE FOSSILS

Fossils of fish, amphibians, reptiles, and dinosaurs are found in rock layers of both parks. In Bryce, there are also fossil bones of crocodiles, turtles, and small mammals. Fossil bones of a plesiosaur, a marine reptile, have been discovered just five miles east of Bryce Canyon.[39] In Zion, dinosaur tracks are found in the Kayenta Formation, a large amphibian fossil was found in the Chinle Formation, and a gar-like fish fossil was discovered in the Moenave Formation.[40]

INVERTEBRATE FOSSILS

Mollusk fossils, including one or more types such as clams, snails, scallops, oysters, and ammonites, are found in the formations in the mile-deep combined layers of Zion and Bryce Canyon National Parks. Snails and clams are so abundant that they are found in the bottom and the top rock layers in both parks, indicating multiple marine environments. The fossilized trails of marine snails or worms are found in the Kayenta Formation in Zion.[40]

DINOSAUR TRACKS

Dinosaur tracks in sedimentary rocks are found by the millions across the earth. A few dinosaur tracks are also found in Zion National Park, for instance near the Human History Museum (see page 36). Evolutionary geologists question how tracks of live dinosaurs could be made while the Flood is raging and depositing a huge volume of sediments.

Actually, it is quite easy.[5] First, as thousands of feet of sediments were laid down early in the Flood, and the area would have become shallow. Then, a short-term fall in sea level (caused by at least four processes) would have briefly exposed the freshly laid sediments. Dinosaurs, either floating in the water or having retreated to higher land nearby, tracked through the freshly exposed Flood sediments, making lots of tracks and even laying eggs. Third, they eventually would be swept up together and die, forming large dinosaur graveyards, when the sea level rose again.

10-inch track of an unknown dinosaur

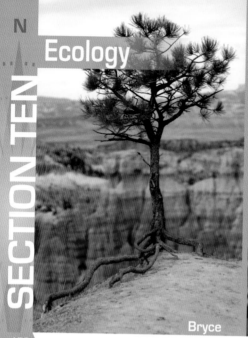

Bryce
Zion

Ecology

A magnificent part of the scenery at Zion and Bryce are the interesting plants and animals. Many are living in delicate balance and interdependence with each other. They have adapted to the rigors of life on the Colorado Plateau where the rainfall is light and undependable and temperatures can be excessively hot in summer and freezing cold in winter. But even with its scarce water and scant soils, these lands are home to thousands of species of plants and animals. A keen observer will have the opportunity to see many of them.

Cone from the bristlecone pine

The study of ecology includes looking at the pattern of where plants and animals live, and why. Ecologists often use the designation "communities" or, more generally, "ecozones" to classify the associations between plants, animals, and their environment. Zion and Bryce can be divided into five broad ecozones:

- Canadian zone – a mix of spruce and fir conifers
- Transition zone – open ponderosa pine forests
- Pinyon-juniper zone – pinyon pine, Utah juniper, and gamble oak
- Upper Sonoran desert zone – sagebrush and cactus
- Riparian zone – area directly dependent upon a permanent water source

It's important to realize that these ecozone distinctions are human inventions designed to help understand the complexities and patterns actually found in nature. But with the unique diversity of environments, plants and animals don't always fit neatly into man's categories. In fact, biologists have found that, historically plant and animal associations are very dynamic and are continually changing over time.

WHAT SHOULD WE EXPECT?

As you consider these life systems, ask yourself which model of origins best explains what you see. The expectations of the two models are shown in the table below, and are far apart in their basic assumptions. As with geology, the conclusions made when examining the ecological environments are influenced by one's worldview (see page 135).

ECOLOGY WORLDVIEW EXPECTATION TABLE

If plants and animals were the result of processes acting over a long period of time, versus the result of a supernatural creation, the following would be expected:	EVOLUTION ▼	CREATION ▼
	long period of time	supernatural creation
Each interbreeding population would produce:	indistinct types as they evolve into new forms	distinct categories, remaining true to their original "kind"
The overall complexity of plants and animals would:	require simultaneous evolution of all parts of a system	be expected and show no signs of transitional forms
The instinct to perform complicated tasks such as migration would:	be communicated or learned through experience	be created and passed on genetically, yet adatable to changing environments
The interdependence that forms cooperative relationships would:	be limited due to competition and the survival instinct	be expected in a creation designed to function as a whole
Creatures (kinds) in different geographical locations would:	show variations due to evolving in different environments	be found essentially the same worldwide

PLANTS AND ANIMALS OF BRYCE AND ZION

In both Bryce and Zion, each slope and fold, each mesa and mountain, and each canyon and valley, creates unique habitats for a variety of plants and animals. For example, although Zion contains only one tenth of one percent of Utah's land area, over 70 percent of the state's native plants are found there. Actually, there is more plant diversity in Zion National Park than on all of the Hawaiian Islands. Interestingly, Zion's canyons are incised into a high plateau and have a unique abundance of water. More than a dozen canyons have perennial streams, which are fed by springs that issue from the bottom of the Navajo Sandstone. This is because the thick, porous sandstone layer acts like a huge stone reservoir. It collects water from the higher elevations that slowly percolates down and out at the lower elevations (see page 51).

Desert spiny lizard

Zion has almost 1,000 species of plants, 95 species of mammals, 40 species of reptiles and amphibians, and over 270 species of birds, 125 of which remain year round. Primarily because of its higher elevations and colder winter temperatures, Bryce has less variety than Zion. Bryce is home to about 250 species of plants, 49 species of mammals, 15 species of reptiles and amphibians, and 170 species of birds, 52 of which are permanent residents.

AMAZING ADAPTATIONS OF DESERT PLANTS

Unlike animals that can move, plants must survive in place and therefore require special mechanisms to find and store water. Some plants have long taproots, which reach deep to find water. For example, young Gamble oak trees may have roots up to seven times deeper than the plant is tall. Others, like cacti and succulents, can quickly grow temporary roots to absorb thunderstorm moisture. The cacti swell to store moisture, which is then used during dry periods.

Obtaining water is one problem; preventing water loss is another. Almost all desert shrubs have small leaves that are waxy or resinous. Many have light gray or hairy leaves to reflect the sun. Manzanita can rotate its leaf edge to reduce the sun's effects (see page 128). Other plants, like juniper, may sacrificially cut off water to an entire branch so the rest of the plant can survive and reproduce. How did these sacrificial survival systems originate?

Desert plants, which grow along streams, are called riparian plants. These are not necessarily drought tolerant but could be called "drought-escaping" plants. These plants often have large leaves, like cottonwoods and maples. Though they can tolerate high temperatures, they are limited to growing along streams and washes. Another group of riparian plants could be called "hanging garden" plants. While there are only a few of these in Bryce, in Zion there are many; and these water-loving plants are found growing at seeps right on the vertical canyon walls.

Pine tree on the Queens Garden Trail in Bryce Canyon

FIRE MANAGEMENT

While driving through the parks, you may notice scorched trees and other signs of recent forest fires. Some fires were started by lightning; others were intentionally set by park staff using an ecological practice called "prescribed burns." For centuries, natural lightning fires crept through these forests every ten to twenty years. These ground fires would renew the forest by burning overcrowded young trees and shrubs as well as accumulations of needle litter and fallen branches.

Prescribed burn in Bryce Canyon

About a century ago, these natural cycles were interrupted by government fire suppression programs. Without regular small fires, forests became overcrowded and more susceptible to big, catastrophic fires, as well as insect and disease attacks. Today, prescribed burn programs are used for the restoration of fire-dependent ecosystems.

ECOLOGICAL EVIDENCES SEEN IN ZION AND BRYCE

The functional complexity and interdependence of living systems points to creation by a Grand Designer. The following four examples are taken from hundreds of possible illustrations seen in Bryce and Zion National Parks. As you consider these amazing creations, ask yourself which model the evidence best supports — evolution or creation?

RAVENS

Raven pair preening

Ravens are possibly the most intelligent birds in the world. Inuit American Indians claim to have used ravens for aerial reconnaissance while hunting. When the raven leads them to a moose, they will reward the birds by leaving some choice cuts of meat behind. Ravens have a "language" of about thirty distinct words or calls with which they communicate, such as sounds like "curruk" or "gluk gluk gluk."

Ravens mate for life and are especially attentive during the breeding season. Pairs are often seen sitting side by side cooing and preening each other. During courtship they often fly together, wingtips touching. Sometimes they perform aerobatics like barrel rolls or flying upside down. Ravens are territorial and will not attempt to produce young if there is not enough food in their territory. Unfortunately, they are attracted to visitors who give them food. The dangers of feeding wildlife also apply to ravens. Watch them around your campsite; they can unzip your pack looking for food!

Bristlecone pine

BRISTLECONE PINE

Because bristlecone pines are thought to be some of the oldest trees in the world, they have achieved great scientific and scenic importance. Unlike other pines, which shed their needles every two to three years, bristlecones hold their needles along the limb for 17 years or more. This causes the needles to stay attached all along the limb, giving the appearance of a small foxtail, hence they are part of a family known as foxtail pines. In Bryce Canyon, they may be seen on the Queen's Garden Trail (left), the Bristlecone Loop Trail at Rainbow Point, along the Fairyland Loop Trail, and scattered at other places, mostly below the rim. One bristlecone pine at Rainbow Point (see page 125) is estimated to be 1,600 years old, which is a youngster compared to the 4,000-year-olds growing in the White Mountains in southeast California.

The study of annual growth rings is called dendrochronology, which claims to date dead trees up to 10,000 years old in the White Mountains of California. These dates present a challenge to the young earth model because a literal biblical chronology places the Flood about 4,500 years ago.

However, research with bristlecone pines grown in a wetter climate found the pines added "an extra ring per annum nearly as often as it will add only a single ring." [41] Unlike popularized ideas, the studies show that the Ice Age was a time of little seasonal contrast with mild winters, cool summers, and heavy precipitation. [13] So the extra annual rings suggesting ages up to 10,000 years old could have been caused by Ice Age climate variations.

YUCCA PLANT AND MOTH

Banana yucca is commonly found in the lower canyon areas of Zion. This versatile plant, sometimes called Datil yucca or Spanish bayonet, has an amazing relationship with a small moth. In fact, the plant and the moth cannot get along without each other. This kind of co-dependency in nature is called biological mutualism.

Yucca moth

This is how it works. The female moth, which only lives for about a week, gathers pollen from the flower, presses it into balls using special mouth parts, and tucks the balls into a pocket beneath her chin. Flying off to another yucca plant, she injects the pollen balls and only one egg into each section of a yucca pod with a special "ovipositor" shaped to do this job. This ensures each section of the flower pod will become fertilized and develop seeds. These seeds will become food for her larva; however, each larva will only eat some of the seeds, allowing the rest to mature.

How did the interdependent body parts of the moth and the yucca match up in the first place? If one of the essential parts in either were missing or if the little moth missed one step, then the whole system would break down. All the interdependent components of this reproductive system had to appear simultaneously. Many scientists, including Darwin himself, doubt the possibility of simultaneous appearance of multiple interdependent parts in living systems. In *The Origin of Species*, Charles Darwin stated: "If it could be demonstrated that any complex organ existed which could not possibly have been formed by numerous, successive, slight modifications, my theory would absolutely break down."[38] Does the complex nature of the yucca and moth reproductive system support Darwin's theory, or provide evidence for its breakdown?

CALIFORNIA CONDORS

California condors are North America's largest bird, with a wingspan up to nine and a half feet. The structure of their wings allows them to soar on rising thermal currents, climbing to altitudes of over 15,000 feet. They often travel over 100 miles a day looking for food and can reach speeds up to 55 mph. They mate for life and can live for 60 years.

Condors once ranged up the west coast of North America and across the south to Florida. In 1805, Lewis and Clark reported seeing them in the Columbia River Gorge, but by 1987 there were only nine wild birds remaining. Through a remarkably successful inter-agency breeding program, they were brought back from the brink of extinction and in 2007 there were over 300 of them. Visitors to Zion and Bryce are likely to spot condors soaring in the skies above. Canyon Overlook in Zion (see page 62) is one of their favorite hangouts.

Condors work cooperatively with the smaller turkey vultures. Turkey vultures have an excellent sense of smell and can find a carcass. A condor's sense of smell is poor, but they have excellent eyesight and while soaring they look for the turkey vultures. The stronger condors can tear open the large carcass, and after eating their fill, leave some for the vultures.

We are told that competition, not cooperation, is the driving mechanism of Darwinian "survival of the fittest" evolution. So how do the many cooperative systems in nature evolve? The answer to that question is simple if your worldview includes a Creator who created them that way "in the beginning."

California condor

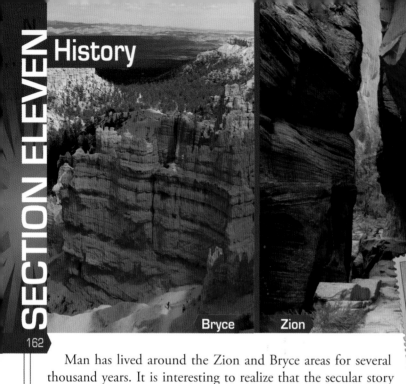

Bryce   **Zion**

1934 Zion
8¢ stamp

Man has lived around the Zion and Bryce areas for several thousand years. It is interesting to realize that the secular story of the migration of man is very similar to the biblical one. With the Ice Age coming on the heels of the Flood,[13] ocean levels would have been lower than today. With more land bridges exposed, the migration of people and animals would have occurred more easily. There is evidence that large Ice Age mammals such as mammoths and giant ground sloths were hunted in southwest Utah near the end of the Ice Age. Following the Ice Age, mobile hunters and gatherers from what is called the Archaic culture roamed the region but left little trace in the archeological record.

Some of the first people to live in the Zion and Bryce areas were the Fremont and Anasazi, also called the Ancient Puebloans. They occupied the territory from about 200 until 1200 A.D. Evidence suggests these two groups traded and intermingled freely. The Anasazi and many of the Fremont people were primarily agriculturalists and mostly lived in permanent settlements. Archaeologists have found pollen of corn, beans, squash, amaranth, and sunflowers in the ruins of these early occupants.

The Anasazi were skilled stonemasons and were also true artisans, making many beautiful items for their households. Cotton, first cultivated by tribes in southern Arizona, was dyed and woven into clothing; yucca and other plant fibers were dyed and woven into sandals and attractive baskets with geometric designs. There is evidence they traded with other groups and learned to make pottery from the Mogollon tribes of the southeast.

Anasazi pot

For nearly 1,000 years, the Ancient Pueblo culture dominated the Colorado Plateau from Zion to the Four Corners area. Then they suddenly left the Bryce and Zion areas around 1200 A.D. A century after that, they disappeared entirely from the Four Corners region, with the last inhabitants living in Mesa Verde and Chaco Canyon. There are many theories regarding their fate. The possibilities supported by archeological evidence include periods of drought, deforestation and soil depletion, hostility from new arrivals, religious and cultural changes, and even influence from the Aztec cultures of Central Mexico. It is believed that the Anasazi are the ancestors of the Hopi, Zuni, and other Pueblo people of the Upper Rio Grande.

Evidence of the ancestral people was left behind on the rock walls of Zion National Park.

Shortly after the Anasazi left, the Southern Paiute people occupied the area. It's interesting that the Paiutes are members of the "Uto-Aztecan" language group, which also included the ancient Mayan and Aztec cultures of Mexico.

The first Europeans to explore the region were two Spanish priests in 1778 named Dominguez and Escalante. They were looking for a route to connect their missions in New Mexico with those in California.

In 1847, Brigham Young led members of the Mormon Church to Utah and settled in the Salt Lake area. Brigham Young chose St. George as the place for construction of the first operating temple of the Church of Jesus Christ of Latter-day Saints (LDS). Mormons came from all over southern Utah to volunteer time, labor, and materials for the temple. Dedicated in 1877, the striking white monument was finished before the better-known Salt Lake Temple. Because it was the first temple in the region and closely associated with the LDS Church founders, it retains a special significance within that church body to this day.

By the early 1900s, the scenic qualities of the region were recognized for its potential for tourism. Yet with the existing roads in poor condition, the area remained inaccessible to most people. In the 1920s, Utah improved the roads, and the Union Pacific extended a spur rail line to Cedar City. From then until now, tourism has steadily increased.

Camping has long been a favorite way to see both parks

ZION NATIONAL PARK

In 1826, a group of mountain men lead by Jedediah Smith explored the region to find a way overland to Spanish California. Known as one of America's greatest explorers, Smith was a Christian and is often portrayed with a rifle in one hand and a Bible in the other. He befriended the Southern Paiutes who led him to Zion Canyon, making him the first "white" man to ever see the canyon.

In the 1860s, John Wesley Powell visited Zion on the first scientific exploration of southern Utah. He later became the first head of the United States Geographical and Geological Survey (now the United States Geological Survey).

Within a decade after they settled in the Salt Lake area, Mormon pioneers were sent out to homestead and grow cotton in "Utah's Dixie." During the 1860s, towns like Grafton and Springdale sprang up, and in 1863, Isaac Behunin homesteaded Zion Canyon near present day Zion Lodge. Isaac was the first to give the name "Zion" to this area and soon it was dotted with other homesteads.

The remainder of the 19th century was a difficult time of change for the Paiutes as well as for the hardy Mormon pioneers. Between floods, sickness, and drought, the small villages and homesteads struggled to survive.

In 1909, President Taft set aside Zion as Mukuntuweap National Monument (mukuntuweap meaning "straight canyon" in Paiute). In 1916, Methodist minister Frederick Vining Fisher visited the canyon. Fisher gave biblical names to many of the canyon's prominent features, including the Watchman, the Altar of Sacrifice, Angels Landing, and the three mountains, which he named after the three patriarchs, Abraham, Isaac, and Jacob. In 1919, it became Zion National Park and celebrated its centennial in 2009.

In 1930, the 1.1 mile long Zion-Mount Carmel tunnel was completed, allowing motorists to travel through Zion to points east. Negotiating the vertical sandstone cliffs of Zion was a great engineering feat and upon its completion, it was the longest tunnel of its kind in the United States. Today, a permit is required for larger vehicles and some motor homes to pass through the tunnel with assistance. Information on restrictions and permits is available on the park's website and in their newspaper, *The Zion Map & Guide*.

To the right is a historic photo of the original Zion Lodge (built in 1925) that burned down in 1966. The rebuilt lodge and other buildings were originally designed by Gilbert Sullivan Underwood. Today, the lodge provides first-class accommodations with a fine restaurant, dining room, café, snack bar, gift shop, and restrooms.

Original 1925 Zion Lodge

BRYCE CANYON NATIONAL PARK

Ebenezer and Mary Bryce

In the 1870s, Brigham Young sent a group to settle the Paria Valley east of Bryce. Among those pioneers were Ebenezer and Mary Bryce. The enterprising Bryce soon built a road to one of the amphitheater canyons to the west, making timber and firewood more accessible. People called it Bryce's Canyon, and although the Bryce family moved to Arizona in 1880, they are still remembered for giving their name to this land.

One of the biggest problems for the Paria Valley pioneers was finding a way to bring more water to the valley. In 1891, their solution was to dig a ditch, by hand, across the top of the Paunsaugunt plateau, and down to the town of Tropic. The ditch was over 15 miles long and the "Tropic Ditch" still supplies irrigation water to the Paria Valley. The ditch crosses under Highway 63 just north of Ruby's Inn on its way to becoming the waterfall near Mossy Cave. An interesting hike is the Mossy Cave Trail that parallels the ditch in this area (see page 25).

In 1916, Ruben (Ruby) C. Syrett brought his family to the wilds of Southern Utah, establishing a cattle ranch near the present site of Ruby's Inn. After seeing the beauty of Bryce Canyon, the family became gracious hosts to help others experience the canyon. What started with tent houses and a place to serve meals paved the way for the modern facilities just outside the park, which now serves thousands of tourists each year.

Bryce Canyon was established as a national monument in 1923 and became Utah National Park in 1924. In 1928, its name was changed to Bryce Canyon National Park.

Bryce Canyon Lodge (below) was built between 1924 and 1925 using local materials and was declared a National Historic Landmark in 1987. The main lodge and over 80 surrounding cabins are tucked away among the pines near Sunset Point.

ADDITIONAL RESOURCES

The following resources provide additional information on subjects based on the same biblical worldview as the True North Series. They do, however, represent a range of theories within that worldview. These resources, along with additional information, are available from the following organizations:

Creation Research Society (CRS), 877 277-2665,
www.creationresearch.org

Answers in Genesis (AiG), 800 778-3390,
www.answersingenesis.org

Institute for Creation Research (ICR), 800 337-0375,
www.icr.org

New Leaf Publishing/Master Books (MB), 800 999-3777,
www.nlpg.com

Guides:

True North Series, Your Guide to the Grand Canyon: A Different Perspective, 2008, T. Vail, M. Oard, J. Hergenrather, and D. Bokovoy, CRS

Road Guide to Yellowstone National Park and Adjacent Areas, 2005, H. Coffin, J. Hergenrather, D. Bokovoy, and M. Oard, CRS

Road Guide to the John Day Area of Central Oregon, 2004, D. Bokovoy, H. Coffin, and J. Hergenrather, CRS

Books:

Ancient Ice Ages or Gigantic Submarine Landslides? 1997, M. Oard, CRS

The Battle for the Beginning, 2001, John MacArthur, AiG

Biology and Creation, 2002, W. Frair, CRS

Bones of Contention, 2004, M. Lubenow, CRS

Earth's Catastrophic Past: Geology, Creation and the Flood, 2009, Snelling, A., ICR

Flood by Design: Receding Water Shapes the Earth's Surface, 2008, M. Oard, CRS

Footprints in the Ash, 2003, J. Morris and S. Austin, ICR

The Fossil Book, 2006, G. and M. Parker, CRS

The Genesis Flood, 1961, J. Whitcomb and H. Morris, ICR

The Geological Column: Perspectives within Diluvial Geology, 2006, edited by J. Reed and M. Oard, CRS

Geology and Creation: 100 Questions and Answers from a Biblical Perspective, 2004, D. DeYoung, CRS

Geology by Design: Interpreting Rocks and the Catastrophic Record, 2007, C. Froede, CRS

The Geology Book, 2000, J. Morris, MB

Grand Canyon: A Different View, 2003, T. Vail, MB

Grand Canyon, Monument to Catastrophe, 1994, S. Austin, ICR

The Great Turning Point, 2004, Terry Mortenson, MB

The Lie: Evolution, 1987, Ken Ham, MB

The Missoula Flood Controversy and the Genesis Flood, 2004, M. Oard, CRS

Natural History in the Christian Worldview, 2001, J. Reed, CRS

The New Answers Book 1, 2006, K. Ham, AiG

The New Answers Book 2, 2008, K. Ham, AiG

The New Answers Book 3, 2010, K. Ham, AiG

The North American Midcontinental Rift System, 2000, J. Reed, CRS

Physical Science and Creation: An Introduction, 1997, D. DeYoung, CRS

Plate Tectonics: A Different View, 2000, edited by J. Reed, CRS

Rock Solid Answers: The Biblical Truth Behind 14 Geological Observations, 2009, edited by M. Oard and J. Reed, CRS

Science and Creation, 2002, W. Frair, CRS

Thousands Not Billions, 2005, Don DeYoung, CRS

The Young Earth Revised & Expanded, 1994, 2007, J. Morris, ICR

DVDs:

Grand Canyon, Monument to The Flood, 1994, S. Austin, ICR

Mount St. Helens, S. Austin, ICR

The Origin of Old-Earth Geology, (Parts 1 & 2), 2003, T. Mortenson, AiG

Rock Strata, Fossils, and the Flood, 2008, A. Snelling, AiG

GLOSSARY

The following definitions of terms refer to their usage in this True North Guide and are not meant to be true "scientific" definitions. Many of these terms also have other meanings when used in a different context.

Alcove — a small, wide recess with an arched overhanging wall of rock

Amphitheater — a steep-sided valley with a wide semicircular upper end

Anticline — a fold in sedimentary rocks that is convex upward

Arch/rock arch — like a natural bridge but without a stream or evidence of a past stream underneath

Basalt — an igneous rock formed from molten lava that normally flows on the surface

Biological mutualism — a biological interaction between individuals of two different species, where both individuals derive a benefit

Brachiopod — a two-shelled marine animal resembling a clam, though generally with its two shells being of different sizes

Carbon-14 dating — a radiometric dating method that uses the naturally occurring isotope carbon-14 to determine the age of carbonaceous unfossilized materials

Channeled Scabland — an area of eastern Washington characterized by vertical-walled canyons and flat bottoms caused by the Lake Missoula flood

Circular reasoning — the logical fallacy in which a premise presupposes the conclusion in some way, yet provides no reason at all to believe its conclusion

Colorado Plateau — a region of plateaus, roughly centered on the Four Corners region (western Colorado, northwestern New Mexico, southeastern Utah, and northern Arizona) which covers an area of approximately 130,000 square miles

Conglomerate — rock consisting of individual water-rounded stones set in a finegrained matrix of sand or silt that has become cemented together

Contact — the boundary surface between two types or ages of rocks; contacts are either "conformable" if they represent no gap of time between them, or an "unconformity" if they are thought to represent a gap of time or erosion between the rocks

Creation — the concept that all humanity, life, and the earth (the universe as a whole) was created by God

Creationism — the belief in a literal interpretation of the Genesis account of creation found in the Bible, which states that God created the universe and all that it contains

Cross-beds — an inclined arrangement of thin sedimentary layers in a larger horizontal layer which indicates the direction of the wind or water currents in the depositional environment

Daughter element — the isotope resulting from the radioactive decay of a parent isotope (e.g., argon gas is the daughter element of potassium)

Dolomite — a sedimentary rock composed largely of calcium magnesium carbonate. Also refers to a mineral, while the rock can be called dolostone.

Ecozone — in ecology, a similar geographic and climate zone usually defined by a few of the dominant plants

Erosion — the displacement of solids (soil, mud, rock, and other particles) by the action of wind, water, or ice

Erosion surface — a geologic feature shaped by the action of erosion

Evolution/Evolutionary (biological) — the belief that all known organisms are related by a common ancestor, having evolved over billions of years by a process in which inherited traits become more or less common in a population over successive generations

Fin — a thin ridge of rock exposed by erosion of vertical cracks that surround the structure

Flood — see Noah's Flood

Formation — a body of rock layers consisting predominantly of a certain type or combination of types of rock

Fossil — the mineralized or otherwise preserved remains or traces (such as footprints) of an animal or plant preserved primarily in sedimentary rock

Geologic column — the column of sedimentary rock that graphically represents the sequence of rock formations for a given locality or region (The entire global geologic column shown in most textbooks is based on a compilation of all the local columns according to the uniformitarian subdivisions of geologic time and is not found anywhere on earth as a continuous sequence.)

Geologic time — a term used by evolutionary geologists to describe the time they believe occurred during earth's history

God — the sole Creator and Sustainer of the universe, as defined in the Holy Bible

Grand Staircase — the area of south central Utah characterized by five cliffs or "stairs" separated by erosion surfaces

Great Denudation — a term referring to the horizontal erosion of great sheets of rock off the Colorado Plateau, especially north of the Grand Canyon, with little if any canyon cutting

Grotto — an alcove caused by sapping or the seeping of water carrying rock particles out of the rock

Hanging valley — a tributary valley that was not eroded down to the level of a main valley

Historical science — the theoretical study of the history of the universe, earth, and its life forms from their origins based on presently observed results; a study not scientifically repeatable, observable, or provable

Hoodoo — a pillar of rock, often forming interesting shapes, left by erosion

Hypothesis — an idea, proposition, or hunch that is tentatively assumed and then tested for validity by comparison with observed facts and by experimentation (less firmly founded than a theory)

Index fossil — a fossil that, in evolutionary thinking, identifies and dates the rock in which it is found, based on when that fossil is assumed to have lived

Layer — rock or soil unit with internally consistent characteristics that distinguishes it from those above and below

Limestone — a sedimentary rock composed largely of calcium carbonate

Liquefaction plume — a vertically moving, generally cylindrical plume of water-carried sediment that moves up under pressure through layers of sediment

Mesa — a steep-sided, flat-topped highland that is smaller than a plateau and larger than a butte

Missing link — see transitional forms

Natural bridge — a bridge of rock formed by a stream that undercuts the rock, usually in a meander

Naturalism/naturalistic — any of several philosophical beliefs that the natural world is the whole of reality and does not distinguish the supernatural from nature

Noah's Flood/The Flood — the global flood described in the Bible that came as a result of God's judgment of sin

Oxidation — the chemical process of combining with oxygen

Paleontologist — a scientist who studies past life forms on earth through the examination of plant and animal fossils

Parent element — the isotope or precursor nuclide from which a daughter element is derived during radioactive decay (e.g., potassium is the parent element of argon gas)

Permeable layer — a layer of rock that allows a flow of water through it

Planation surface — a land surface shaped by the action of erosion, especially by running water, and generally applies to a flat or nearly flat planed off surface

Radioactive decay — the set of various processes by which an unstable atom (parent element) decays to produce a different atom (daughter element)

Radioisotope dating — a technique used to date metamorphic and igneous rocks based on the assumed decay rate of the specific elements and the resulting ratio of parent and daughter elements found in the rock today

Riparian — the ecosystem of plants and animals living along a water source which are dependent on that water source

Sandstone — a sedimentary rock composed mainly of sand-sized mineral or rock grains cemented together

Sapping — the process in which groundwater exits a bank or hillside laterally in the form of a seep or spring, eroding soil from the slope and often causing the collapse of material above

Scripture — the Bible, a collection of 66 books considered by Christians to be inerrant and literally inspired by God; also referred to as the Holy Bible or Word of God

Sedimentary rock — a layered rock formed by the accumulation and consolidation of sediments

Shale — a fine-grained sedimentary rock whose original constituents were clay, mud, and/or fine silt

Sheet erosion — the erosion of material by water flowing over land as a widespread mass instead of in definite channels or rills

Sin (as used in the Bible) — to miss the mark and fall short of the perfection of God

Slick rock — areas of barren rock, normally sandstone, which has been eroded smooth, but not necessarily flat, such as the area surrounding Checkerboard Mesa in Zion National Park

Slot canyon — a narrow, vertically walled canyon

Speciation — the development of new species from existing ones

Theory — a concept or proposition developed and better substantiated than a hypothesis but not so conclusively proven as to be accepted as a law

Transitional form/fossil — the fossilized remains of a life form that illustrates an evolutionary transition; a human construct that vividly represents a particular evolutionary stage, as recognized in hindsight

Travertine — a coating or buildup of limestone formed by the evaporation of water

Unconformity — a buried erosion surface separating two rock masses or layers of different ages, indicating that sediment deposition was not continuous; in the evolutionary model this also represents a lengthy interval of missing time

Underfit stream — a stream that appears too small to have eroded the valley or canyon in which it flows and too small to remove the associated rock debris

Uniformitarian/Uniformitarianism — the philosophy that assumes the natural processes operating in the past were the same as those observed operating in the present, often summarized in the statement, "The present is the key to the past"

Weathering — the process of breaking down rocks, soils, and their minerals through direct or indirect contact with the atmosphere and moisture, which, contrary to erosion, occurs without movement of the particles

Window arch — a small arch (in Bryce easily eroded by weathering)

Word, the — see Scripture

Worldview — the framework through which an individual interprets the world and interacts in it

REFERENCES

1. Mortenson, T., 2004, *The Great Turning Point*, Master Books, Green Forest, AR.

2. Leet, D., 1982, *Physical Geology*, Prentice-Hall, Englewood Cliffs, NJ.

3. Vail, T., 2003, *Grand Canyon, a Different View*, Master Books, Green Forest, AR.

4. Twidale, C. R., 1968, *Geomorphology*, Thomas Nelson, Melbourne, Australia, p. 164-165.

5. Oard, M. J., 2009. Dinosaur tracks, eggs, and bonebeds, In, Oard, M. J. and J. K. Reed (editors), *Rock Solid Answers: The Biblical Truth Behind 14 Geological Questions*, Master Books and Creation Research Society Books, Green Forest, AR and Chino Valley, AZ, pp. 245-258.

6. Gitt, W., 1994, *If Animals Could Talk*, CLV, Bielefeld, Germany, p. 75.

7. Anonymous, 2001, *The Zion Canyon Shuttle Guide*, Zion Natural History Association, Springdale, UT, p. 18.

8. Eves, R. L., 2005, *Water, Rock & Time*, Zion Natural History Association, Springdale, UT, p. 116.

9. Austin, S. A., 1994. Interpreting strata of Grand Canyon. In, Austin, S. A. (editor), *Grand Canyon: Monument to Catastrophe*. Institute for Creation Research, Dalles, TX, pp. 32-35.

10. Baars, D. L., 2000, *The Geology of the Colorado Plateau: A Geologic History*, revised and updated, University of New Mexico Press, Albuquerque, NM.

11. Rahl, J. M., P. W. Reiners, I. H. Campbell, S. Nicolescu, and C. M. Allen, 2003. Combined single-grain (U-Th)/He and U/Pb dating of detrital zircons from the Navajo Sandstone, Utah. *Geology* 31(9): 761-764.

12. Freeman, W. E. and G. S. Visher, 1975. Stratigraphic analysis of the Navajo Sandstone. *Journal of Sedimentary Petrology* 45(3), p. 651.

13. Oard, M. J., 2004. *Frozen In Time: The Woolly Mammoth, the Ice Age, and the Biblical Key to Their Secrets*, Master Books, Green Forest, AR.

14. Harris, A. G., E. Tuttle, and S. D. Tuttle, 1997. *Geology of National Parks*, Kendall/Hunt Publishing Co., Dubuque, IA, p. 49.

15. Brady, I., 2004. *The Redrock Canyon Explorer*, Nature Works Press, Talent, OR.

16. Minta, et al., 1992, Hunting associations between badgers and coyotes, *Journal of Mammalogy* 73:814-820.

17. Crickmay, C. H., 1974. *The Work of the River: A Critical Study of the Central Aspects of Geomorphology*, American Elsevier Publishing Co., New York, NY.

18. Oard, M. J., 2008, *Flood by Design: Receding Water Shapes the Earth's Surface*, Master Books, Green Forest, AR.

19. Perloff, J., 1999. *Tornado in a Junkyard: The Relentless Myth of Darwinism*. Refuge Books, Burlington, MA.

20. Ham, K., *The Lie: Evolution*, Master Books, Green Forest, AR

21. Powell, J. L., 2005. *Grand Canyon: Solving Earth's Grandest Puzzle*, Pi Press, New York, NY, p. 219.

22. Walker, T., 1994. A Biblical geological model. In, Walsh, R. E. (editor), *Proceedings of the Third International Conference on Creationism*, technical symposium sessions, Creation Science Fellowship, Pittsburgh, PA, p. 581-592.

23. Oard, M. J., 2004, *The Missoula Flood Controversy and the Genesis Flood*, Creation Research Society Books, Chino Valley, AZ.

24. Morales, M., 1990, Mesozoic and Cenozoic strata of the Colorado Plateau near the Grand Canyon. In, Beus S. S. and M. Morales (editors), *Grand Canyon Geology*, Oxford University Press, New York, NY, p. 247-260.

25. Schmidt, K., 1989, *The significance of scarp retreat for Cenozoic landform evolution on the Colorado Plateau, U.S.A.* Earth Surface Processes and Landforms, British Society for Geomorphology, 14:93-105

26. Vail, T., M. Oard, D. Bokovoy, and J. Hergenrather, 2008. *Your Guide to the Grand Canyon: A Different Perspective*, Master Books, Green Forest, AR.

27. Rigby, J. K., 1977. *Southern Colorado Plateau*, K/H Geology Field Guide Series, Kendall/Hunt Publishing company, Dubuque, IA, p. 3.

28. Kocurek, G. and R. H. Dott, Jr., 1981. Distinctions and uses of stratification types in the interpretation of eolian sand. *Journal of Sedimentary Petrology* 51: 579-595.

29. Pazzaglia, F. J., 2004. Landscape evolution models. In, Gillespie, A. R., S. C. Porter, and B. F. Atwater (editors), *The Quaternary Period in the United States*, Elsevier, New York, NY, p. 249.

30. Twidale, Ref. 4, pp. 148-203.

31. Brand, L. 1997, *Faith, Reason, and Earth History*, Andrews University Press, Berrien Springs, MI.

32. Biek, R. F., G. C. Willis, M. D. Hylland, and H. H. Doeling, 2003. Geology of Zion National Park, Utah. In, Sprinkel, D. A., T. C. Chidsey, Jr., and P. B. Anderson (editors), *Geology of Utah's Parks and Monuments*, Utah Geological Association Publication 28, second edition, Salt Lake City, UT, p. 107-137.

33. Roth, A. A., 1998. *Origins—Linking Science and Scripture*, Review and Herald Publishing, Hagerstown, MD.

34. Bates, R. L. and J. A. Jackson (editors), 1984. *Dictionary of Geological Terms*, third edition, Anchor Press/Doubleday, Garden City, NY, p. 170.

35. Bates and Jackson, Ref. 34, p. 387.

36. Dury, G. H., 1977. Underfit streams: retrospect, perspect, and prospect. In, Gregory, K. J. (editor), *River Channel Changes*, John Wiley & Sons, New York, NY, p. 281-293.

37. DeYoung, D., 2005, *Thousands…Not Billions: Challenging an Icon of Evolution Questioning the Age of the Earth*, Master Books, Green Forest, AR.

38. Darwin, C., 1859. *The Origin of Species*, John Murray, London, UK.

39. DeCourten, F., 1994. *Shadows of Time, The Geology of Bryce Canyon National Park*, Bryce Canyon Natural History Association, p. 42-57.

40. Hamilton, W. L., 1978 (revised 1987) *Geological Map of Zion National Park, Utah*, Zion Natural History Association, p. 76–87.

41. Lammerts, W. E., 1983. Are the bristle-cone pine trees really so old? *Creation Research Society Quarterly* 20(2): 108-115.

ILLUSTRATIONS & PHOTO CREDITS
A= All, T=Top, M=Middle, B=Bottom, L=Left, R=Right

ILLUSTRATIONS:
Jennifer Bauer: 119B
Bill Looney: 102B
Dan Stelzer: 142-143T

PHOTOS:
All photography by Tom Vail unless otherwise noted.
Brian Tuten: 70B
Cory Lawler: 122B
Dave Welling: 100B
Dennis Bokovoy: 36B, 14B, 67MR, 71-73B, 111-112B, 152B, 155A
Ian Parker; Evanescent Light Photography: 103-104B
Istockphoto: 10 (Gray Fox), 26MR, 39TL, 54B, 56B, 68B, 84B, 89B, 95M, 98B,
    107B, 111M, 114B (Douglas fir), 118B, 119ML, 120B, 133B, 162M, 163
John Hergenrather: 38BL, 88B, 127M
Julian Nicholson: 94-97T
Lee Dittmann: 86B
Leon Tuten: 99M
Mike Oard: 61B, 67BL, 90B, 130B, 149M
National Park Service: 12T, 14TL, 20, 25TL, 27B, 31, 75, 76MR, 85Ml, 85MR,
    85BL, 85BR, 91BL, 106B, 147B, 163B
Photos.com: 41B, 183MR
Renee Yelton: 137B
Robert Mitchell: 160B
Shutterstock: Front Cover, 1, 2, 3, 12B, 18B, 19B, 24TR, 24BL, 25BR, 40BR, 43B
    (smoke), 47T, 48B, 58B, 59B, 72M, 72ML, 87B, 96BR, 99B, 101B, 104BR,
    106T, 108-109T, 108B, 113B, 115T, 126B, 131M, 131B, 134TR, 134M,
    135BL, 135BR, 136B, 138A, 152Tl, 152TR, 156TR, 156TL, 157B, 162TL,
    162TR
United States Geological Society: 46B
Wikipedia: 95BL, 114B (Limber pine & white fir)

INDEX

ABOUT THE AUTHORS

Dennis Bokovoy followed his childhood interests in geology and fossils, earning a MS degree in geology from Montana State University. His 30 years as an educator include 15 years as an Adjunct Professor of Geology. Dennis and his wife, Yvonne, live in Hood River, Oregon. He and John Hergenrather are part of a field team co-authoring road guides from a creation perspective, including Yellowstone National Park and the John Day area of Central Oregon. In addition, Dennis is a tour leader for Creation Encounter Ministries (www.creationencounter.com).

John Hergenrather graduated from Oregon State University with a bachelor's degree in geography. John is Vice President of the Design Science Association, a Portland, Oregon-based creation science group. For 15 years, he has been part of a creation geology research team which co-authored road guides from a creation perspective. John and his wife, Rhea, live in Hood River, Oregon. Besides researching and writing creation guidebooks to the National Parks, John conducts field trips through Creation Encounter Ministries (www.creationencounter.com).

Michael Oard is a retired meteorologist from the National Weather Service and an amateur geologist. He has spent more than 30 years studying the earth sciences from a creation perspective, writing many books and articles. He is a popular speaker and has authored many creation-based books including: *The Weather Book, Frozen in Time, The Missoula Flood Controversy and the Genesis Flood,* and *Flood by Design.* Mike and his wife, Beverly, live in Bozeman, Montana, and together they co-authored a children's book titled *Life in the Great Ice Age.* Mike was also involved as a co-author in the *Yellowstone National Park Road Guide.*

Tom Vail started guiding trips in the Grand Canyon in 1980 and is the author of *Grand Canyon, a Different View,* which provides a creation perspective of the Grand Canyon. Tom and his wife, Paula, live in Phoenix, Arizona, and together run Canyon Ministries, providing Christ-centered rafting trips through the Grand Canyon. Once an evolutionist, Tom now welcomes opportunities to share the biblical message of the Grand Canyon. You can find out more about their ministry at *www.CanyonMinistries.com.*